A Practical Guide to Power Line Communications

This excellent resource synthesizes the theory and practice of power line communication (PLC), providing a straightforward introduction to the fundamentals of PLC as well as an exhaustive review of the performance, evaluation, and security of heterogeneous networks that combine PLC with other means of communications. It advances the groundwork on PLC, a tool with the potential to boost the performance of local networks, and provides useful worked problems on, for example, PLC protocol optimization. Covering the PHY and MAC layers of the most popular PLC specifications, including tutorials and experimental frameworks, and featuring many examples of real-world applications and performance, it is ideal for university researchers and professional engineers designing and maintaining PLC or hybrid devices and networks.

Christina Vlachou is Senior Research Scientist at Hewlett Packard Enterprise Laboratories.

Sébastien Henri is Senior Software Engineer at Cisco Meraki.

A Practical Guide to Power Line Communications

CHRISTINA VLACHOU

Hewlett Packard Enterprise Laboratories

SÉBASTIEN HENRI

Cisco Meraki

CAMBRIDGE
UNIVERSITY PRESS

University Printing House, Cambridge CB2 8BS, United Kingdom

One Liberty Plaza, 20th Floor, New York, NY 10006, USA

477 Williamstown Road, Port Melbourne, VIC 3207, Australia

314–321, 3rd Floor, Plot 3, Splendor Forum, Jasola District Centre, New Delhi – 110025, India

79 Anson Road, #06–04/06, Singapore 079906

Cambridge University Press is part of the University of Cambridge.

It furthers the University's mission by disseminating knowledge in the pursuit of education, learning, and research at the highest international levels of excellence.

www.cambridge.org
Information on this title: www.cambridge.org/9781108835480
DOI: 10.1017/9781108890823

© Christina Vlachou and Sébastien Henri 2022

First published 2022

A catalogue record for this publication is available from the British Library.

ISBN 978-1-108-83548-0 Hardback

Contents

Preface

Having spent eight years on research of power line communications (PLC), we observed a significant gap between industry and academia, especially on the configuration and management of PLC devices. From the industry perspective, there are a few open source tools for configuration and measurement of PLC. However, unlike Wi-Fi chipsets, there is no open source firmware. Measurement and configuration of commercial devices require a lot of reverse engineering. From the academia perspective, there are many analytical and measurement works, which often do not get compared to or implemented on commercial devices due to the tools' limitations and obfuscated firmware. During our doctoral theses, we developed an experimental framework on commercial devices as well as analysis tools for PLC. Performance analysis is useful for network optimizations and for ensuring scalability in multiuser deployments. Yet, models have to be validated against experimental results, and chipset configuration is essential for implementing any optimizations. To help users and researchers address these issues, we have decided to write this book to share our knowledge and methods to configure, manage, analyze, and evaluate PLC networks.

Our book mainly focuses on the IEEE 1901 standard, on which the vast majority of commodity PLC devices are based. We first provide an understanding of the standard, giving the most important features of physical (PHY) and medium-access control (MAC) layers. Understanding how data flows and is modulated and transmitted over the electrical wires is crucial for evaluating PLC performance. The standard is about 1600 pages and describes two types of PLC networks: the Internet access networks and the indoor enterprise/residential PLC networks. Hence IEEE 1901 stations can be deployed in-building or over power line distribution cables. The two deployments differ in topology, protocols, and channel quality, which differentiates also the PHY and MAC features. In our book, we focus on broadband indoor PLC networks, which work at low voltage and are very popular in residential environments.

The book is divided into three parts. First, we present the channel models and PHY layer of PLC devices. We explain how data flows are modulated into analog signals transmitted over the electrical wires and how these signals propagate. The attenuation and noise experienced by PLC signals are different and more complex than those of Wi-Fi. This yields a different PHY layer design with adaptive and periodic modulation with respect to the alternate current (AC). The MAC layer of PLC is also more complex than Wi-Fi. Similarly to Wi-Fi, PLC is a broadcast medium; hence stations have to resolve collisions when they transmit simultaneously. But owing to the specificities of power lines, PLC MAC introduces an additional variable that creates different levels of multiuser efficiency and short-term fairness. We present a detailed experimental framework for configuring and measuring performance of PLC commercial devices. We discuss how to measure PHY- and MAC-layer metrics, configure topology and security mechanisms, and write new tools for these functionalities.

In the second part of the book, we present a performance evaluation of the PHY and MAC layers on a testbed of nineteen stations. We explore the spatial and temporal variation of capacity and provide guidelines for link metric estimation in hybrid PLC/Wi-Fi networks. Then, we present efficiency and throughput models of the MAC layer and analyze the multiuser performance and configurations. In this second part, we provide experimental guidelines to reproduce the experiments.

In the third part of the book, we discuss security, management, and other applications of PLC. We present the processes of creating secure PLC networks, new station authentication and association, and potential security attacks on PLC. Finally, we discuss the IEEE 1905.1 standard for heterogeneous networks where several technologies, such as PLC and Wi-Fi, coexist. We detail the most important features of heterogeneous networks. We conclude the book by providing new research directions and open problems of PLC. Throughout the book, we compare Wi-Fi and PLC, as they are technologies often used simultaneously in heterogeneous networks. The PHY and MAC layers of PLC have differences that can benefit the network in terms of coverage, throughput, latency, and security.

The book can be useful for PLC researchers, users who would like to optimize PLC performance, engineers working on new PLC devices, and computer networking classes. Our book is intended for both a general audience with little networking background and an advanced audience with a Wi-Fi, PLC, or networking background. To distinguish the parts of the book for the advanced audience, we introduce sections for further reading using the 👀 symbol. The book is also intended to be used as a guide for PLC testbed development, configuration, and measurement. When applicable, we provide experimental guidelines to reproduce our results or network configurations using the 💻 symbol.

This book is a product of eight years of research on PLC; a significant part of this research was carried out under the supervision of Professor Patrick Thiran and Professor Albert Banchs. We thank them for their guidance and advice on Chapters 5 and 6. Patrick supervised our theses and helped us with the analysis and performance evaluation of PLC. Albert has helped us with the throughput model of the PLC MAC layer. The performance evaluation of PLC was done at École polytechnique fédérale de Lausanne (EPFL), Switzerland, on a testbed that Julien Herzen had initially developed for Wi-Fi networks. We thank Julien for his help with the testbed and the inauguration of PLC devices on it. Finally, we thank Can Karakuş for his feedback on the book.

We hope that the book will serve as a practical guide on teaching PLC, new research directions on PLC, and development of new PLC commercial devices.

Abbreviations

AC alternating current
ACK acknowledgment
AES advanced encryption standard
AFE analog front end
AGC automatic gain control
ALME abstraction layer management entity
AP access point
ARP address resolution protocol
ARQ automatic repeat request
ASCII American standard code for information interchange
BBF bidirectional burst flag
BBT beacon backoff time
BC backoff counter
BDF beacon detect flag
BIFS burst interframe space
BLE bit loading estimate
bps bits per second
Bps bytes per second
BPSK binary phase-shift keying
CA channel access
CBC cypher block chaining
CCo central coordinator
CFP contention-free period
CFS contention-free session
CIFS contention interframe space
CMDU control message data unit
CMG CTS-MPDU gap
CP contention period
CRC cyclic redundancy check
CSC channel-switching cost
CSMA/CA carrier sense multiple access with collision avoidance
CTS clear to send

CW contention window
D/A digital-to-analog converter
DAK device access key
dB decibel
dBm decibel of measured power referenced to one milliwatt
DC deferral counter
DPLL digital phase-locked loop
DPW device password
DSNA device-based security network association
DTEI destination terminal equipment identifier
EIFS extended interframe space
EKS encryption key select
EMI electromagnetic interference
EPFL École polytechnique fédérale de Lausanne
ETH Ethernet
ETT expected transmission time
ETX expected transmission count
FC frame control
FCCS frame control check sequence
FDM frequency division multiplexing
FEC forward error correction
FFT fast Fourier transform
FID fragment identifier
FL frame length
Gbps gigabits per second
G.hn specification for home networking
GHz gigahertz
GI guard interval
GP GreenPHY
GUI graphical user interface
HD high definition
HD-PLC high-definition power line communication alliance
HF high frequency
HLE higher layer (above MAC) entity
HPAV HomePlug AV
HPGP HomePlug Green PHY
Hz hertz
IARU International Amateur Radio Union
ICV integrity check value
IEEE Institute of Electrical and Electronics Engineers
IFFT inverse fast Fourier transform
IFS interframe space
IMC impedance mismatch compensation
IoT Internet of Things

IP Internet protocol
IPTV IP television
ISI intersymbol interference
ISP intersystem protocol
ITU International Telecommunication Union
ITU-T ITU Telecommunication Standardization Sector
IV initialization vector
kbps kilobits per second
KCD key carrying device
kHz kilohertz
LAN local area network
LDPC low-density parity check codes
LID link identifier
LLDP link layer discovery protocol
LN logical network
LTE long-term evolution 4G mobile communications standard
m meters
MAC medium-access control
Mbps megabits per second
MCF multicast flag
MCS modulation and coding scheme
MHz megahertz
MID message identifier
MIMO multiple input and multiple output
MLME MAC layer management entity
MM management message
MNBC multinetwork broadcast
MNBF multinetwork broadcast flag
MoCA Multimedia over Coax Alliance
MPDU MAC protocol data unit
MPTCP transport control protocol
MRTFL maximum reverse transmission frame length
ms millisecond
µs microsecond
MSDU MAC service data unit
NEK network encryption key
NFCNK near-field communication network key
NID network identifier
NMK network membership key
NMK-SC network membership key – simple connect
NPW network password
NVRAM nonvolatile random access memory
OFDM orthogonal frequency-division multiplexing
OFDMA orthogonal frequency-division multiple access

PB physical block
PBC push-button configuration
PBCS physical block check sequence
PBKDF password-based key derivation function
PC personal computer
PCP priority code point
PE protective earth
PHY physical layer
PIB parameter information block
PKCS public-key cryptography standard
PLC power line communications
PLME physical layer management entity
PPB pending physical block
PPDU physical protocol data unit
PRS priority resolution slot
PSD power spectral density
QAM quadrature amplitude modulation
QoS quality of service
QPSK quadrature phase shift keying
QUIC Quick UDP Internet connections
RCG RTS/CTS gap
RF radio frequency
RI roll-off interval
RIFS response interframe space
ROBO robust modulation schemes for IEEE 1901
RSC recursive systematic convolutional
RSNA robust security network association
RSSI received signal strength indicator
RTS request to send
RTT round trip time
s second
SACK selective acknowledgment
SC simple connect
SHA secure hash algorithm
SIFS short interframe space
SISO single input and single output
SL security level
SME station management entity
SNID short network identifier
SNR signal over noise ratio
SoF start of frame
SSID service set identifier
SSN sequence segment number
STEI source terminal equipment identifier

SVD single-value decomposition
SWM sliding-window method
TCC turbo convolutional coder
TCP transport control protocol
TDM time division multiplexing
TDMA time division multiple access
TEI terminal equipment identifier
TEK temporary encryption key
TIA Telecommunications Industry Association
TLV type length value
TTL time-to-live
TV television
UCPK user-configured passphrase
UDP user datagram protocol
UIS user interface station
UKE unicast key exchange
UPA Universal Powerline Association
USA United States of America
USAI unassociated station advertisement interval
VLAN virtual local area network
VLC visible light communication
VoIP Voice over IP
WiMAX worldwide interoperability for microwave access
WPA Wi-Fi protected access
WPS Wi-Fi protected setup
WSC Wi-Fi simple configuration
XOR exclusive or

1 Introduction

1.1 Motivation and Goals

Enterprise and residential buildings are facing an explosion in connectivity demands for high-bandwidth applications and automation. Networking infrastructure needs to be scaled to satisfy such demands and to provide reliable, full coverage for myriad devices and applications. Although wireless communications are pervasive and their efficiency has been largely improved by recent amendments to IEEE 802.11 standard, wireless networks often face coverage and congestion problems due to a limited unlicensed spectrum, thus failing to guarantee high quality of service to the end user. Nowadays, we observe an explosion of connectivity demands in all kinds of environments, from residential and enterprise to rural and urban, which is a challenge for state-of-the-art networking solutions. In addition, many applications have multiple requirements from communication technologies, such as mobility, ease of use, performance, and reliability. Performance can often be measured by throughput, which is the number of data bits delivered to the device by the network per second, or latency, which is the total time required to transmit data (e.g., a packet) from the moment it enters the transmission queue. Both throughput and latency depend on multiple network layers and are typically measured and characterized as the quality of service (QoS) of the application.

Wireless communication is dominant in residential and enterprise networks; it offers mobility and attractive data rates. Nevertheless, it often leaves "blind spots" in coverage due to building architecture. Wireless signals can be attenuated by walls built of brick or stucco, which act as an obstacle to network access for certain users. A study in 1000 American residences suggests that 40 percent of them have experienced problems with the wireless routers in their networks [1]. Among those households experiencing connectivity problems, 63 percent continue to have these problems despite some efforts to resolve them. One common solution is to incorporate wireless repeaters and to enable multihop communication. Yet, wireless signals are still obstructed by construction obstacles, often failing to extend coverage reliably.

Owing to the shared nature of wireless media, another connectivity issue arises from neighboring networks that cause interference to each other. This is usually addressed by assigning a frequency channel that is noninterfering with neighboring networks. However, channel allocation is becoming increasingly challenging, first, because of the density of wireless networks in urban environments, and second, because of recent Wi-Fi standardizations, such as IEEE 802.11ac/ax, that maximize utilized bandwidth (e.g., to 80/160 MHz), resulting in very few available noninterfering channels.

In addition to wireless signal attenuation and interference, there can be connectivity disadvantages with wired technologies like Ethernet. For instance, smart static devices, such as TVs or network drives, often enable only Ethernet connectivity, which requires costly infrastructure. Even when static devices offer Wi-Fi capabilities, users often prefer wired connections for nonmobile applications to avoid wasting Wi-Fi bandwidth. Compared to Wi-Fi, Ethernet usually does not have good coverage, and the ease of use is limited due to long wiring and high cost.

Wireless networking problems are exacerbated by congestion created by the number of connected devices. In recent years, a vast number of mobile devices has been added in everyday life: smartphones and tablets; wearables; and automation devices for comfort, energy savings, and security. In fact, Cisco estimates that by 2030, we will have more than 500 billion connected devices [2]. The explosion in the number of connected devices per building and the Internet of Things (IoT) networking require communication technologies and protocols that can scale and that react efficiently to congestion.

Beyond IoT and smart devices, wide bandwidth and decreases in processing costs enable many bandwidth-hungry applications, such as augmented reality, ultra high-definition streaming, gaming, video and web conferencing, cloud storage, and virtual desktops. Under increased bandwidth demands, Wi-Fi might fail to meet the required quality of service in case of multiple applications or contending flows, despite the recent amendments for high efficiency. These amendments have pushed most Wi-Fi parameters to the capacity limit (e.g., channel bandwidth, modulation, number of antennas and spatial streams). Hence, in the future, Wi-Fi capacity improvements will be limited. Moreover, as mentioned earlier, wired Ethernet technologies – often the most popular substitute for Wi-Fi – require costly infrastructure, confining their ubiquitous coverage. For all these reasons, additional communication mediums are required for bandwidth-hungry applications.

The aforementioned connectivity problems and demands call for hybrid networks combining multiple efficient and user-friendly communication technologies. The incorporation of additional communication technologies to today's networks can bring multiple benefits. In addition to bandwidth aggregation, diverse mediums can augment network reliability, extend coverage, and provide different levels of quality of service in the network. Multiple technologies enable an exploitation of medium diversity and accommodation of more users and applications. This will enable the user to eliminate connectivity problems and meet high rate demands, in addition to resilience, a reliable backbone, and security.

In this book, we explore a promising communication technology for hybrid residential and enterprise networks: power line communications (PLC). PLC is beneficial in terms of both easy installation and high data rates: it enables data transmission via the electrical wires and through common power outlets – a ubiquitous infrastructure; it is trivial to install, as it does not require new wiring; and it provides rates up to 1.5 Gbps. Most common applications of PLC include Wi-Fi coverage extension and replacement of costly Ethernet connections, as depicted in Figure 1.1.

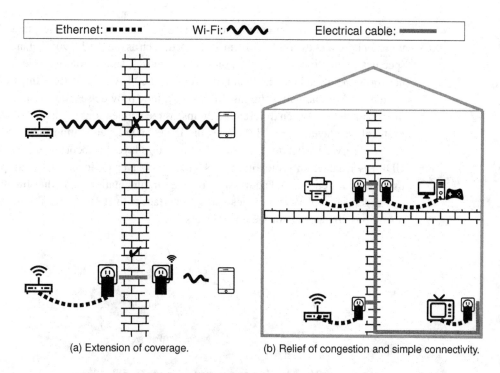

Figure 1.1 Examples of PLC addressing connectivity problems and demands. The icons of this book have been made by Freepik from www.flaticon.com.

The first goal of this book is to show the fundamentals of PLC in terms of channel characteristics, physical (PHY) and medium access control (MAC) layer functions, security, and performance. Second, the book provides guidelines to configure PLC devices, measure statistics, ensure security in data transmissions, and troubleshoot performance. To this end, we present a performance evaluation using the guidelines introduced in this book, pointing out the advantages and challenges of PLC, of which users and administrators should be aware.

In the remainder of this chapter, we first explore the evolution of hybrid networks and the candidate technologies for augmenting network reliability. We then turn our attention to power line communications. We discuss high-bandwidth and IoT PLC applications, standardizations, and specifications. Finally, we present the book's organization.

1.2 Local Networking Technologies

We discuss the state-of-the-art efforts for combining multiple communication technologies into heterogeneous (or hybrid – the two terms are used interchangeably) networks. Wi-Fi is always included in hybrid solutions, as it offers mobility and serves the vast majority of mobile applications. We explore candidate technologies that have the potential to augment Wi-Fi reliability, and we discuss their benefits and drawbacks.

As the demand for combining noninterfering technologies increases, new specifications for hybrid networks have been developed, such as the IEEE 1905.1 standard, which specifies abstraction layers for topology, link metrics, and forwarding rules [3]. Hybrid networks operate at layer 2.5, that is between IP and MAC layers, according to standardization and commercial solutions. In this way, they provide seamless connectivity and interoperability between different mediums, as the applications work above the Internet Protocol (IP) layer. Figure 1.2 shows the IEEE 1905 layers within the TCP/IP stack. As we observe, applications run on top of transport layer protocols, which are typically UDP (user datagram protocol) or TCP (transport layer protocol). The transport layer uses the IP for end-to-end addressing, routing, fragmentation, and other functions. The convergent layer of 1905 provides a single interface to the IP layer, resulting in seamless operations, irrespective of the lower layers.

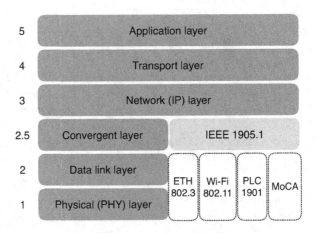

Figure 1.2 IEEE 1905 layer in the TCP/IP stack.

IEEE 1905.1 provides a common interface for widely deployed home networking technologies, such as wireless, power line, coaxial, and Ethernet communications. PLC, Wi-Fi, Ethernet, and coaxial (MoCA) technologies are defined and operate on layers 1 and 2 (physical and data link layers) of the TCP/IP stack, lower than the convergent layer of IEEE 1905, as we see in Figure 1.2. The MAC layer is one of two sublayers of the data link layer in the TCP/IP stack (the other sublayer is called logical link control). Multiple vendors have released hybrid products combining PLC and Wi-Fi or coaxial communications and Wi-Fi. In addition to reliability, full coverage, and bandwidth aggregation, IEEE 1905.1 targets energy management, simple and push-button security, and advanced network diagnostics. We describe the IEEE 1905.1 standard in Chapter 8.

Several candidates for enhancing Wi-Fi performance are on the market, among which are power line and coaxial communications. We now review these technologies in the quest for reliable, cost-effective, and easy-to-use communications. Table 1.1 summarizes the advantages and disadvantages of four popular communication technologies. We discuss each technology separately.

Table 1.1 IEEE 1905.1 candidate technologies for hybrid networks

Technology		Advantages	Disadvantages
Wi-Fi	📶	Mobility and ease of use	• Prone to interference • Walls cause severe attenuation
PLC	🔌	High availability of outlets and ease of use	• Prone to interference • Performance varies depending on wiring quality and electrical grid structure
MoCA	◉	High performance and reliability	Requires wiring and infrastructure
Ethernet	▥	High performance and reliability	Requires new wiring and infrastructure

Wi-Fi is undoubtedly the most popular solution due to the explosion of mobile applications. The other candidate technologies often act as coverage extenders for Wi-Fi. The main reasons for the existence of these extenders are the connectivity demands and problems discussed in Section 1.1. In recent years, multichannel Wi-Fi is widely employed, with multiple Wi-Fi devices simultaneously utilizing channels in the 2.4 GHz and 5 GHz bands (such routers are called dual-band). The advantage of such solutions is mainly bandwidth aggregation; coverage extension and resilience are more challenging when using only two Wi-Fi bands, as all wireless band capacities are affected by similar correlated factors (e.g., signal penetration, mobility, fading). 60 GHz Wi-Fi is also being developed and offers multi-Gbps capacities, but its short range and blockage due to the human body prevent 60 GHz Wi-Fi from being a candidate for coverage extension.

PLC is becoming very popular in home networks. The main advantage of this technology is the no-new-wires connectivity and the high density of electrical plugs in any residential or enterprise environment. Furthermore, PLC is a natural and trivial solution for smart electrical appliances already connected to the grid. The potential disadvantages of PLC are interference – because PLC is a shared medium like Wi-Fi– and high performance variability depending on the age of the electrical grid or the electrical appliances operating, which affect the channel quality. Because of the similarities between wireless and power line media, many PHY/MAC layer techniques from Wi-Fi can be reused for PLC.

Coaxial communication is specified and designed by the Multimedia over Coax Alliance (MoCA) [4], and it is usually known simply as MoCA. Communication over coaxial cables achieves 1.4 Gbps data rates, using bands in the frequency range of 500–1650 MHz and with 100 MHz bandwidth. The main advantage of this technology is high reliability with packet error probability of about 10^{-6}. The interference is also limited to within the local network, as shielded coaxial wires prevent interference from neighboring households or other technologies. MoCA applications mainly focus on multiroom digital video recorders, over-the-top streaming content, gaming, and ultra high-definition (HD) streaming. Although it provides high reliability and rates,

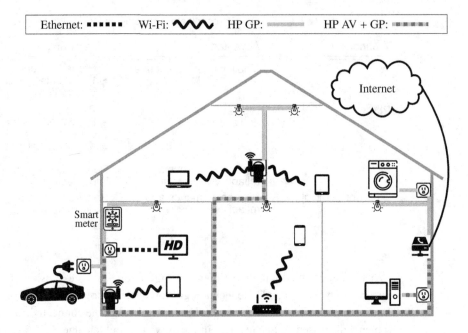

Figure 1.3 Connected home with PLC technologies: HomePlug AV/AV2/AV500 (HP AV) and HomePlug GreenPHY (HP GP).

MoCA requires an already established coaxial system and possibly more than two coaxial plugs, which are limited in today's households.

Finally, Ethernet communication provides the highest reliability, but it can be costly. It is usually deployed in enterprise buildings with a sophisticated network structure. In residential environments, users often employ Ethernet for nonmobile applications, and its main disadvantage is the requirement for new wiring. Therefore, in comparison to MoCA and Ethernet, PLC is a more promising technology for hybrid networks in terms of cost, flexibility, and pervasiveness of connectivity possibilities.

1.3   Power Line Communication: Applications and Market

1.3.1   Easy, Low-Cost, High-Throughput Connectivity

Owing to the growing demand for reliability in home networks, wireless and power line communications are combined by several vendors to deliver high rates and broad coverage without blind spots. PLC is at the forefront of hybrid networking for residences, as it provides easy, plug-n-play, and high-data-rate connectivity. Its main advantages are its wider coverage compared to Wi-Fi, data rates up to 1.5 Gbps, and the fact that it does not require new wiring or infrastructure.

PLC can handle bandwidth-hungry applications thanks to its high data rates, but it can also be employed for home automation and low-rate networking. Smart appliances

(TV, refrigerator, lights, etc.) are inherently connected to the power grid, hence they can use the power line for both power supply and communications. This property of PLC proves very efficient for future connected homes, as they will use multiple smart appliances for comfort, security, and energy savings. Figure 1.3 presents examples of residential PLC applications.

Because of all the advantages discussed above, PLC is widely adopted in residential and enterprise networks. HomePlug, the leading alliance for PLC standardization and certification, estimates that more than 180 million PLC devices have already been shipped and that the expected annual growth is more than 31 percent over the next years. HomePlug also proposes different solutions for home automation and high-data-rate local networks. The most popular specification for high-data-rate PLC, employed by 95 percent of PLC devices, is HomePlug AV. This specification was adopted by the IEEE 1901 standard [5].

In this book, we focus on indoor and broadband PLC, in the frequency range of 1–80 MHz, and more specifically on the HomePlug AV and IEEE 1901 specifications. However, note that there also exist narrowband solutions for outdoor reliable communications with applications on power distribution automation, demand-response control, and meter-to-grid connectivity. These solutions are implemented in rural, urban, and suburban areas. The IEEE 1901.2 standard provides guidelines for narrowband PLC using frequencies below 500 kHz and data rates of up to 500 kbps [6].

1.3.2 PLC for Bandwidth-Hungry Applications

PLC has been very popular in home networking for extending Wi-Fi coverage and for providing connectivity to a wide range of applications. PLC vendors offer a wide range of devices supporting different data rates and Wi-Fi capabilities. In many residential environments, PLC solves Wi-Fi connectivity issues thanks to hybrid devices that provide multihop PLC/Wi-Fi connectivity to mobile users in different rooms and floors. Figure 1.3 shows examples of such hybrid devices. In general, PLC can have a much wider coverage – up to 300 m – than Wi-Fi. However, as we discuss in this book, performance largely depends on the power grid structure.

PLC is also being used as an additional medium for aggregating bandwidth along with Wi-Fi. Consumers of PLC devices usually connect PLC to TV, gaming consoles, whole-home audio, security cameras, and network storage. They exploit the abundant power outlets and electrical wire infrastructure in residential buildings. PLC promises Gbps rates, and hence can support 4K/HD streaming, file sharing, and other demanding multimedia applications.

PLC has been considered a reliable medium for in-vehicle communication (e.g., car, plane, train, and ship scenarios) [7]. For instance, PLC can exploit the existing power distribution infrastructure to deliver high-speed services to every cabin or to every seat in vehicles. Such applications are becoming popular, but their deployment requires further research.

Finally, PLC has use cases in enterprise environments too. Offering additional bandwidth, it can be used for fault tolerance, coverage extension, and smart offices. Recently,

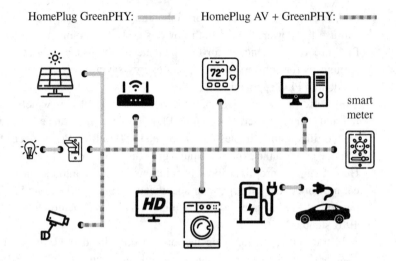

Figure 1.4 HomePlug GreenPHY examples.

PLC has been shown to efficiently work on data centers [8], for managing multiple servers and adding resilience to a system where communications could fail for multiple reasons.

1.3.3   PLC for Home Automation and IoT

PLC is an ideal networking candidate for smart appliances connected to the power grid. HomePlug alliance released the specification GreenPHY for applications that require low-power, low-rate, reliable networking. GreenPHY has been defined and developed in collaboration with the US Department of Energy, utility companies, and automobile manufacturers. One of the useful features of GreenPHY is the interoperability with high-bandwidth AV devices, hence potentially with the entire home network, without the need for special bridges to convert communications signals, as we show in Figure 1.4.

Multiple solutions use GreenPHY, ranging from home automation to vehicle charging. Smart appliances using GreenPHY, such as refrigerators, washers, heating, and air conditioning, contribute to an energy-efficient and connected home. GreenPHY is also integrated as a communication solution for electrical vehicle charging [9]. GreenPHY enables robust, low-power, and low-cost communications for all these applications.

HomePlug GreenPHY chipsets have more flexibility in deployment compared to AV. The chipsets can be integrated in many sensor systems and hardware platforms. For example, there are guidelines for integrating GreenPHY chipsets with Rasberry Pi [10]. This gives a wide range of possibilities and applications and enables users to create custom automation systems for residential or enterprise environments.

In addition to HomePlug GreenPHY, there are other popular communication protocols for home automation and smart metering over power lines. For instance, INSTEON uses the X10 technology developed in 1975 on 131.65 kHz channels [11, 12]. X10 is limited to specific commands (e.g., commands that switch on/off a device, switch all units off, turn all lights off), whereas GreenPHY enables generic low-rate communications. Similarly, Netricity, which is built on the IEEE 1901.2 standard, enables 500 kbps rates on frequencies below 500 kHz [13]. Applications of Netricity span over smart grid, distribution automation, and net metering demand and response.

1.4 Standardizations and Specifications

Broadband PLC specification started developing in the late 1990s, mainly supported by the HomePlug Powerline Alliance, Universal Powerline association (UPA), and High-Definition Power Line Communication (HD-PLC) alliance. The timeline of PLC specifications is shown in Table 1.2. The first specification, HomePlug 1.0, was released in 2001 and supported 14 Mbps capacity. The Telecommunications Industry Association (TIA) published the international standard TIA-1113 in 2008, which was based on HomePlug 1.0, released 7 years earlier. HomePlug AV is the successor of HomePlug 1.0, published in 2005. HomePlug AV substantially improves capacity upon earlier specifications, yielding up to 200 Mbps. Since its inception, HomePlug AV has led the PLC market with hundreds of millions of devices.

HD-PLC and HomePlug had developed different protocols for PLC. The existence of noninteroperable PLC devices initiated a standardization effort by both IEEE and ITU-T, in 2005 and 2006, respectively. The IEEE 1901 and ITU-T G.9960/61 standards for PLC were released in 2010, including coexistence protocols for diverse devices [14, 15]. Coexistence is enabled through the intersystem protocol (ISP) specified in IEEE 1901 and ITU-T G.9972, whose support is mandatory for IEEE 1901 devices [16].

IEEE 1901 defines the protocols and functions of PHY/MAC layers for high-speed power line communications. The standard specifies protocols for both in-home networking and access networking, in the frequency range 1.8–50 MHz. Access is defined as a communications system of high-speed data on frequencies above 2 MHz over medium- and low-voltage power lines that distribute electricity to end user premises, enabling a communication link between the power utility infrastructure and the residence or enterprise. The power lines can be outside or within a building. The IEEE 1901.2010 standard consolidated two existing specifications in the market: fast Fourier transform (FFT) orthogonal frequency-division multiplexing (OFDM) modulation from HomePlug and TIA-1113 technologies and wavelet OFDM modulation from HD-PLC technology. HomePlug AV and GreenPHY are interoperable with IEEE 1901. Green-PHY was introduced in 2010, mainly for reliable and low-rate communications, as it only uses the most robust modulation schemes of IEEE 1901. It has up to 75 percent lower power consumption compared to HomePlug AV. The amendment IEEE 1901a in 2019 defines a flexible way of separating wavelet OFDM channels for IoT networks.

IEEE has published the **IEEE 1901.2** standard for narrowband communication, which targets communication below 500 kHz and smart grid applications [6]. In 2018, IEEE also published **1901.1** standard for medium-frequency (less than 12 MHz) PLC

Table 1.2 PLC standardization timeline

	HomePlug	ITU-T	IEEE
2001	HomePlug 1.0 release		
2005	HomePlug AV release		IEEE 1901: project authorization
2006		ITU G.hn: project authorization	
2008	TIA-1113 release		IEEE 1901 technical proposal selection
2010	HomePlug GreenPHY release	G.9960, G.9961, G.9972 published	IEEE 1901 standard release
2011		G.9963 published	
2012	HomePlug AV2 release		
2013			IEEE 1901.2 and 1905.1 release
2014			IEEE 1905.1a amendment
2018			IEEE 1901.1 standard
2019			IEEE1901a amendment

and smart grid applications [17]. Finally, the **IEEE 1905.1** standard focuses on heterogeneous networks comprising Ethernet, PLC, Wi-Fi, and MoCA.

The **ITU-T G.hn** standard defines networking over power lines, phone lines, and coax cables with data rates up to 1 Gbps, unlike IEEE 1901, which is restricted to power lines. G.hn is a home network standard, as opposed to being just a PLC standard, and targets combinations of existing wiring in a typical home (such as phone, power, and coax cables) to deliver a unified network. G.9960/61 standards define the PHY and data link layers for communication, and G.9972 defines coexistence functions between G.hn and IEEE 1901 with ISP. Synchronization with ISP is achieved using a special OFDM signal of sixteen continuous symbols.[1]

After IEEE 1901 and ITU-T G.hn were published, new specifications were released for Gbps rates. For instance, HomePlug AV2 optimizes PHY and MAC layers for 1.5 Gbps throughput, while maintaining interoperability with IEEE 1901 and older HomePlug devices. Compared to older specifications, state-of-the-art standardizations reduce overhead in the MAC layer and introduce multiple antenna techniques (MIMO) by exploiting the existing two-wire electrical infrastructure for additional antennas. HomePlug AV2 and ITU-T G.9963 use MIMO transmissions on the PHY layer [18].

---

[1] A symbol is a waveform for a fixed period of time.

1.5   Book Organization

In this book, we focus on the IEEE 1901 standard and HomePlug specifications.

Part I. How Does PLC Work?

Chapter 2 We overview the main characteristics of the power line channel, such as noise, attenuation, and its broadcast nature. We identify the key factors that affect end-to-end performance of single links. We discuss the PHY layer functions and the evolution of PLC technologies. We present a typical PLC transceiver, its signal modulation and coding techniques, and their parameters. We discuss the new features of HomePlug AV2 compared to IEEE 1901 and the differences between Wi-Fi and PLC PHY layers.

Chapter 3 We then turn our attention to efficiency when multiple users contend for the medium. To resolve contention conflicts, PLC uses carrier sense multiple access with collision avoidance (CSMA/CA) on the MAC layer. The stations have to sense the medium, and wait for a random interval of idle-medium time slots, before they transmit. The PLC CSMA/CA protocol is similar but more complex than that of Wi-Fi. We present the IEEE 1901 CSMA/CA protocol and certain MAC-layer processes, such as the priority resolution for QoS classes, interframe spaces, and frame aggregation. We discuss the new features of HomePlug AV2 compared to IEEE 1901 and the differences between Wi-Fi and PLC MAC layers.

Chapter 4 We introduce an experimental framework for PLC. We explain how to configure PLC devices and measure certain statistics, such as the capacity of the links, packet errors, modulation information, and collision statistics. We rely on PLC management messages and on open source tools. We give examples of these messages and guidelines on employing the tools. We also explain how to develop new custom PLC tools.

Part II. How Does PLC Perform?

Chapter 5 We move to our experimental study on the end-to-end capacity of PLC links. We discuss the variability of PLC capacity over space, frequency, and time by using a testbed on a realistic enterprise environment. These measurements are used for characterizing end user performance, for establishing practical link-metric guidelines, and for facilitating future deployments of hybrid networks that comprise PLC technologies. We present link-metric guidelines that use metrics proposed by the IEEE 1905.1 standard.

Chapter 6 We evaluate the performance of the PLC CSMA/CA protocol by analysis, simulations, and testbed measurements. We explain the impact of each parameter on performance and the motivation behind the design choices for PLC. Both throughput and short-term fairness are discussed and modeled. Short-term fairness is achieved when all users share the medium equally over a short time horizon. If short-term fairness is not guaranteed, there exists high delay deviation which can harm certain delay-sensitive applications. We provide fundamental evidence about performance trade-offs of the PLC CSMA/CA. Our study suggests that PLC parameters should be judiciously selected to exploit the trade-off between throughput and fairness and to achieve high QoS for best-effort and delay-sensitive applications. This chapter also presents a cross-layer PHY/MAC evaluation of contending links and investigates the challenges of stations with different rates and hidden terminals.

Part III. Management, Security, and Further Applications

Chapter 7 We explain the fundamentals of PLC network management and security. We describe how privacy and security are guaranteed in PLC. We introduce guidelines for configuring security keys in commercial PLC devices. Finally, we discuss potential issues with PLC security and compare PLC with Wi-Fi from a physical layer and security protocol perspective.

Chapter 8 We give details on the IEEE 1905.1 standard for hybrid networks. The standard employs, among others, Wi-Fi, PLC, Ethernet, and MoCA technologies. We cover the main components of the standard and connect to our performance evaluation and guidelines in previous chapters.

Chapter 9 We give concluding remarks on our performance evaluation of PLC. We underline the main challenges for PLC and its future integration to enterprise networks. We present some new research directions.

References

[1] Parks Associates for Multimedia over Coax Alliance, "Consumer Interest in Wired Solutions to Wireless Home Networking Problems: Quantitative Findings," Tech. Rep., 2015, [Online]. Available: www.mocalliance.org/news/Parks-Associates-Final-Report.pdf. [Accessed: November 6, 2019].

[2] Cisco, "Internet of Things. Connected Means Informed." Tech. Rep., July 2016, [Online]. Available: www.cisco.com/c/dam/en/us/products/collateral/se/internet-of-things/at-a-glance-c45-731471.pdf. [Accessed: November 6, 2019].

[3] *IEEE Standard for a Convergent Digital Home Network for Heterogeneous Technologies*, IEEE Standard 1905.1-2013.

[4] Multimedia over Coax Alliance (MoCA). [Online]. Available: www.mocalliance.org/. [Accessed: November 6, 2019].

[5] *IEEE Standard for Broadband over Power Line Networks: Medium Access Control and Physical Layer Specifications*, IEEE Standard 1901-2010.

[6] *IEEE Standard for Low-Frequency (Less than 500 kHz) Narrowband Power Line Communications for Smart Grid Applications*, IEEE Standard 1901.2-2013.

[7] M. Antoniali, A. M. Tonello, M. Lenardon, and A. Qualizza, "Measurements and analysis of PLC channels in a cruise ship," in *2011 IEEE International Symposium on Power Line Communications and Its Applications (ISPLC)*, pp. 102–107.

[8] L. Chen, J. Xia, B. Yi, and K. Chen, "Powerman: An out-of-band management network for datacenters using power line communication," in *Proceedings of the 15th USENIX Symposium on Networked Systems Design and Implementation (NSDI)*, 2018, pp. 561–578.

[9] J. Zyren, "Electrical Vehicle Charging with HP GreenPHY," Qualcomm, Tech. Rep., Apr. 2015, [Online]. Available: www.qualcomm.com/sites/ember/files/uploads/ev_combined_charging_qualcommautotechconf_april_2015.pdf. [Accessed: November 6, 2019].

[10] M. Heimpold, "Using Raspberry Pi with HomePlug Green PHY Powerline Evaluation Kit," I2SE GmbH, Tech. Rep., Jan. 2017, [Online]. Available: download.i2se.com/ApplicationNotes/an6_rev1.pdf. [Accessed: November 6, 2019].

[11] INSTEON. [Online]. Available: www.insteon.com/technology#ourtechnology. [Accessed: November 6, 2019].

[12] X10 Knowledge base. [Online]. Available: kbase.x10.com/wiki/Main_Page. [Accessed: November 6, 2019].

[13] Netricity. [Online]. Available: www.netricity.org/. [Accessed: November 6, 2019].

[14] *Unified High-Speed Wireline-Based Home Networking Transceivers – System Architecture and Physical Layer Specification*, IITU-T G.9960.

[15] *Unified High-Speed Wireline-Based Home Networking Transceivers – Data Link Layer Specification*, IITU-T G.9961.

[16] *Coexistence Mechanism for Wireline Home Networking Transceivers*, IITU-T G.9972.

[17] *IEEE Standard for Medium Frequency (Less than 12 MHz) Power Line Communications for Smart Grid Applications*, IEEE Standard 1901.1-2018.

[18] *Unified High-Speed Wireline-Based Home Networking Transceivers – Multiple Input/Multiple Output Specification*, IITU-T G.9963.

Part I

How Does PLC Work?

2 The PHY Layer of PLC

2.1 Introduction

In this chapter, we discuss the main characteristics of PLC channel and PHY layer functions to understand how they affect end user performance. In addition, we build a solid background for the following chapters of this book.

A power line is a very harsh channel in terms of noise and attenuation. Noise consists of unwanted external signals in the medium. It is often characterized by its power spectral density (PSD), which is the magnitude squared of the Fourier transform of a signal. Attenuation is the reduction of the amplitude of the original signal at the receiver and is usually measured in decibels (dB). Compared to the wireless medium, the power line has diverse types of noise that cannot be modeled as Gaussian, which is the typical wireless noise distribution. Moreover, it exhibits multipath effects and high-frequency selectivity. We review the PLC channel characteristics and the factors that degrade PLC performance in Section 2.2. We explain how communication is achieved over the wires and the broadcast nature of electrical wires. Then, in Section 2.3, we present the most popular specifications and standardizations that succeed in mitigating these harsh channel conditions and in providing promising data rates.

The complexity of PLC channel properties leads to a more sophisticated design for both PHY and MAC layers compared to Wi-Fi. We give an overview of the PHY layer of PLC in Section 2.4, where we discuss the main PHY features of a PLC transceiver. In Section 2.5, we describe the modulation and coding schemes, and in Section 2.6, we distinguish the differences between HomePlug AV and AV2 specifications. In Section 2.7, we uncover how these schemes are chosen per link, that is, the channel estimation and adaptation processes. Section 2.8 gives IEEE 1901 parameter values and timing. Section 2.9 discusses the differences between the IEEE 1901 and ITU-T G.hn standards. Section 2.10 compares PLC and Wi-Fi on the PHY layer. This study is crucial for understanding end-to-end performance and for fully using PLC in hybrid networks later in Chapters 5 and 8.

2.2 Channel Characteristics

PLC signals are transmitted via electrical wires between outlets. Hence they "share" the channel with multiple electrical appliances and grid components (such as circuit breakers and distribution lines) that create a harsh environment for communications. We explain the main components of the channel, that is, attenuation and noise, which affect both the spatial (across different outlets) and temporal (for a certain pair of outlets, over time) variations of PLC capacity. Consider an example of a simple electrical network with a transmitter (TX) and a receiver (RX), as given in Figure 2.1. The main sources of attenuation and noise are the electrical appliances plugged in. Modeled with dashed boxes in Figure 2.1, each connected appliance has an impedance and produces a noise process that is non-Gaussian and that depends on the device type [1].

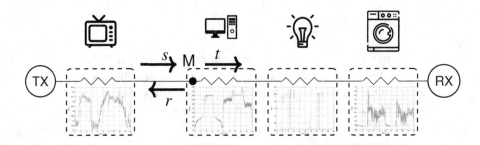

Figure 2.1 Multipath and noise in PLC channels. We show the noise measured by prior studies, used with permission from IEEE (2010). *Time frequency analysis of noise generated by electrical loads in PLC* [1], © 2010 IEEE. All rights reserved.

2.2.1   Attenuation and Frequency Selectivity

With respect to communications, the electrical cable becomes a transmission lin with a characteristic impedance. The connection of appliances creates impedance mismatches to this transmission line, causing the transmitted signal to be reflected multiple times. The signal reflection amplitudes at impedance mismatches depend on the type and working mode of the appliances, that is, on/off and idle/active. For example, in Figure 2.1, at point M, we have an impedance mismatch, and any signal s arriving at M is partly reflected (signal r) and partly propagates (signal t) in the same direction as the original signal s. Reflections of signals at various impedance-mismatched points result in multiple versions of the initially transmitted signal arriving at different times at the receiver, thus establishing a multipath channel for PLC.[1]

Further Reading To explain impedance mismatch effects, we consider the PLC cable as a transmission line. Every transmission line (cable) has a characteristic impedance Z_0 that depends on the cable's circuit coefficients and operating frequency:

$$Z_0(\omega) = \sqrt{\frac{R + j\omega L}{G + j\omega C}},\qquad(2.1)$$

---

[1] Wireless channels are also multipath due to signal reflections in the environment (e.g., walls, objects, people).

where ω is the angular frequency and R, L, G and C are the per-unit-length resistance, inductance, conductance, and capacitance of the transmission line, respectively. For a coaxial cable, typical values of Z_0 are 50 or 75 Ω. For PLC frequencies, it has been shown that $R \ll \omega L$ and $G \ll \omega C$. Hence the characteristic impedance can be approximated by $Z_0 \sim \sqrt{L/C}$. With this approximation, Z_0 does not depend on the frequency that PLC uses (e.g., 1–50 MHz) [2]. In typical transmission lines, impedance Z_0 must match the input impedance of the transmission line load to ensure maximum power transfer over the line. Most transmission lines incur some load Z_L due to transmitter/receiver devices or other imperfections. In PLC, the load is due to the electrical appliances attached and any other grid components with impedance. In the general case of a load Z_L, the impedance measured at a given distance l from the load is given by

$$Z_l = Z_0 \frac{1 + \Gamma_L e^{-2\gamma l}}{1 - \Gamma_L e^{-2\gamma l}}, \tag{2.2}$$

where γ is a propagation constant and $\Gamma_L = (Z_L - Z_0)/(Z_L + Z_0)$ is a reflection coefficient. The equations above are derived by using the so-called telegraph equations, which are linear differential equations describing the voltage and current on an electrical transmission line. In PLC, owing to the high diversity and number of electrical appliances attached, we have multiple reflection points with diverse reflection coefficients. Hence a PLC channel is challenging to model because exhaustive measurements of the impedance Z_L of multiple electrical appliances are needed. A common approach for modeling the channel is to derive a channel response function for a multipath of N paths and estimate the coefficients numerically as a function of frequency f [2]:

$$H(f) = \sum_{i=0}^{N} \underbrace{g_i}_{\text{weighting factor}} \underbrace{e^{-(\alpha_0 + a_1 f^k)}}_{\text{attenuation}} \underbrace{e^{-j2\pi f(\sqrt{\epsilon_r} d_i / c_0)}}_{\text{delay}}. \tag{2.3}$$

The weighting factor g_i is a complex number and depends on the frequency; it is a product of transmission and reflection coefficients due to impedance mismatches. The more transitions and reflections occur along a path, the smaller the weighting factor. Prior measurement studies have shown that it is possible to simplify the weighting factors by real, frequency-dependent numbers. In such a case, the weighting factor represents the weight of a path in the multipath channel. The attenuation factor parameters a_0 and a_1 measure the cable losses, and k is a real number with $0.5 \le k \le 1$. Again, these parameters can only be empirically measured. Finally, the delay factor represents the different arrivals of the paths based on the distance of the cable d_i for path i, the speed of light c_0, and the dielectric constant ϵ_r.

Attenuation is determined by the distance between TX and RX due to cable losses and by the reflections of the signal. The delay spread – that is, the delay difference between the arrival of strong reflections at the receiver – of the PLC channel is typically in the order of 1–2 μs, with a few exceptions over 5 μs [3]. The PHY layer of PLC is designed to mitigate this delay spread and these multipath effects, as we will see in Section 2.4. Multipath effects cause intersymbol interference and frequency selectivity. They induce more severe attenuation than potential losses due to long cable lengths.

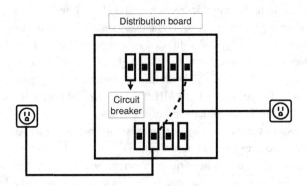

Figure 2.2 Example of two PLC devices connected on different circuit breakers.

Frequency selectivity may cause the signal to experience different, uncorrelated attenuation levels based on frequency. The spatial variation of PLC channel is mainly dictated by the position, the impedance, and the number of appliances connected to the electrical grid, rather than by the cable length between the stations – although the number of appliances and the cable length can be correlated. Since the aggregate electrical activity in residential or enterprise environments varies during the day, the channel transfer function and attenuation change at large timescales (order of minutes or hours). Hence temporal variation of the PLC channel is caused by the electrical activity, that is, the appliances' impedances, on long-term timescales, whereas it is caused mainly by noise on short-term timescales, as we discuss in Section 2.2.2.

The level of attenuation depends on the number of reflection points, also called branches. It has been found that links of distance up to 100–200 m and with up to four branches mostly exhibit typical attenuation values starting from a few decibels at 500 kHz and reaching 40–70 dB at 20 MHz frequency [2]. Longer links (>300 m) exhibit 10–30 dB at 500 kHz and might exceed 80 dB at frequencies of 5–8 MHz (which corresponds to the noise floor).

2.2.1.1   Additional Attenuation Factors

In addition to attenuation introduced by electrical appliances, there are attenuation factors that are usually not considered in channel modeling and that arise due to the structure of the electrical grid and power distribution. For instance, significant attenuation can be introduced between a pair of stations connected to different circuit breakers, current phases, or distribution boards. Figure 2.2 presents an example of two PLC stations connected through different circuit breakers. PLC devices can be connected via the same cable on a specific circuit breaker, different circuit breakers, or even distribution boards. The impedance of the appliances and these extra attenuation factors render the prediction of PLC link quality challenging, given the location of two stations. In Chapter 5, we discuss these factors, and we observe in our testbed that spatial variation of PLC is difficult to predict.

2.2.2 Types of Noise

Various types of noise are generated in a PLC channel. A widely accepted classification of PLC noise (e.g., [1, 4, 5]) identifies the following types.

Colored background noise is due to different low-power noise sources present in the network and is usually characterized with a power spectral density decreasing with frequency. This noise can be modeled by a colored Gaussian power spectral density.

Narrowband interferences are caused by radio broadcasters and are usually sinusoidal signals with modulated amplitudes (due to radio modulations).

Impulsive noise is caused by electrical appliances: they create synchronous or asynchronous noise to AC mains and aperiodic noise when switched on/off. It is further distinguished into three classes that we describe next. Usually, the impulses in PLC channels are modeled with damped sine waves.

The root mean square (rms) amplitudes of the colored background noise and narrowband interference vary slowly over time, and they can be classified as background noise. Impulsive noise has high variability and is the most severe and intricate type of noise, because each noise source (electrical appliance) exhibits unique characteristics, from both time and frequency perspectives. In detail, the three classes of impulsive noises are as follows.

Periodic, synchronous to the mains noise is a cyclostationary noise, synchronous with the mains and with a frequency of 50/60 Hz (depending on the country). It is often generated by rectifier diodes used in electrical appliances. Examples of such noise are given in Figure 2.1.

Periodic, asynchronous to the mains noise is generated mostly by switched power supplies and has a frequency of tens or hundreds of kilohertz (in the range 12–217 kHz [5]).

Aperiodic noise is caused by people switching electrical appliances on and off and plugging them in.

The short-term variation of the channel is due to noise. Two timescales describing this variation, called invariance and cycle scales, are introduced by Sancha et al. [6]. An invariance scale is defined at a level of subintervals of the mains cycle and is associated with the noise synchronous to mains, whereas the cycle scale represents different mains cycles and is caused by noise asynchronous to mains. PLC modulation and coding are tailored to the noise variation within or across the mains cycle, as we explain in Section 2.4.

ⵎ Further Reading Noise modeling for PLC is challenging, and several studies have focused on developing statistical models for the different types of noise described above [1, 7].

Colored background noise can be modeled (according to the IEEE 1901 standard) with a power spectral density (PSD) at frequency f, $S_{CBN}(f) = -a + b|f|^{-c}$ dB-m/Hz, where all parameters a, b, c are positive. The parameter values are estimated empirically and given by the standard. In addition, the standard highlights that another PSD $S_{CBN}(f) = a + be^{-f/f_0}$, where f_0 is a reference sequence, is also validated experimentally. Empirical values for these parameters in a residential environment are $a = b = 35$, $f_0 = 3.6$ [8].

Narrowband interference noise (the noise due to narrowband interferers) has been modeled with the PSD $S_{NI} = \sum_{k=0}^{N} A_k \cdot e^{-(f-f_{0,k})^2/2 \cdot B_k^2}$ [8], where we have N interferers with center frequency $f_{0,k}$, bandwidth B_k, and amplitude A_k. The parameter values can in turn be modeled by different distributions depending on the assumptions on the interferers' types and locations.

Periodic noise can be modeled based on the observation that the impulses in PLC channels very often consist of damped sinusoids,[2] $p(t) = \sum_{i=0}^{N_I-1} A_i e^{-\alpha_i t} e^{-j2\pi f_i t}$, where N_I is the number of impulses and damped sinusoids, A_i is a zero-mean complex Gaussian variable independent of i, and α_i are damping factors that control the corresponding pulse width [7]. The periodic impulsive noise can be written as a function of the sum of periodic impulses $n_{PI}(t) = \sum_n p_n(t - nT)$, where T is the period of n_{PI}. With experimental data, we can find typical T values and other parameters at different times of day and in different environments (countries, residential or enterprise scenarios).

Aperiodic noise is produced by human activity introducing or removing electrical appliances, and this type of noise is modeled by a Poisson process with average arriving rate λ, describing the appliances' switching transients. Because human activity varies over the day, λ modeling changes at different times of day. In Chapter 5, we observe in practice how capacity varies over 24 hours and across days. Measurement studies have highlighted that aperiodic impulses resemble a damping sinusoidal waveform. Each random impulse can be expressed as $p(t) = \sum_{i=0}^{N_I-1} A_i e^{-\alpha_i t} e^{-j2\pi f_i t} u(t)$, where N_I is the number of impulses and damped sinusoids, A_i is a zero-mean complex Gaussian variable independent of i, α_i are damping factors that control the corresponding pulse width, and $u(t)$ is the unit-step function [7].[3] The random impulsive noise can be written as a function of the random impulses $n_{RI}(t) = \sum_n p_n(t - t_n)$, where the arrival difference $t_n - t_{n-1}$ is an exponential random variable independent for different ns, with expectation $1/\lambda$.

The total noise at the PLC receiver depends on the channel transfer function. The PSD is the sum of background noise and the noise produced by appliances $S(f) = S_{CBN} + S_{NI} + S_{appliances}$, where $S_{appliances}$ is the summation of noise power spectrum from all the appliances $S_{appliances} = \sum_k S_k(f)|H_k(f)|^2$ and $H_k(f)$ is the transfer function from the noise source k location to the receiver. In any model and PLC simulation, a researcher has to assume or measure certain distributions for the random variables representing the numbers, locations, and types of appliances. 🔭

---

[2] A damped sine wave is a sinusoidal function whose amplitude approaches zero as time increases.

[3] The definition of the unit step function: $u(t) = 1$, if $t \geq 0$, and $u(t) = 0$, if $t < 0$.

2.2.3 PLC: A Broadcast Medium

PLC devices connected to the same distribution board form a broadcast network, where almost all stations can overhear each other's transmissions. To understand the reasons for such good connectivity, consider the power distribution network: it forms a star topology where power flows from the board to any circuit breaker, and then to power outlets. From a circuit breaker to the wall outlets, the wiring has a tree topology. Any cable can form a communication "antenna." Hence any signal from an outlet could potentially reach any other outlet connected to the same board. In Chapter 5, our experiments show that PLC usually forms almost a fully connected network on a single distribution board. In addition, we find that connectivity between devices connected to different distribution boards within the same building is also possible. Such links suffer from severe attenuation due to long cable length and additional impedance introduced by board components.

The extremely good connectivity and the star like topology of PLC result in the tragedy of the commons. Similarly to Wi-Fi, which is also a broadcast medium, PLC stations cannot transmit simultaneously. Collision detection (detection of interference during transmission) is currently not supported in PLC due to near-far effects – where the self-interference of the device drowns out weaker interfering signals – and to the lack of appropriate interference cancellation techniques. Hence collision avoidance protocols are needed in the presence of multiple transmissions. We explain these protocols in Chapter 3.

2.2.4 Electromagnetic Interference (EMI)

In addition to causing interference to other PLC stations, the broadcast nature and radiation of PLC can result in electromagnetic interference to other communication technologies. Electrical power is distributed at 50/60 Hz frequency, whereas PLC operates at frequencies above 1 MHz. Hence EMI caused by PLC could harm AM and amateur radio bands. Radiation comes from the unshielded power line wires. However, PLC radiation is local and usually confined within the building. To avoid such radiation, IEEE 1901 and other PLC standards require frequency notches that attenuate specific PLC frequency chunks. A notch filter is a band-stop filter with a very narrow band rejection. Figure 2.3 shows the spectral mask and notch filters as defined by IEEE 1901 over 1–30 MHz. The maximum transmission power of an IEEE 1901 station is −55 dBm/Hz. The notches mainly protect amateur radio bands by attenuating frequencies in the radio range to a very low level (−85 dBm/Hz).

We present two major studies that elaborate on EMI caused by PLC [9, 10]. Ofcom [9] explores the likelihood and extent of interference caused by PLC. The study was initiated in 2010 due to complaints of interference caused by PLC in the UK, mainly by radio enthusiasts. This work uses simulations and data sets from realistic applications, such as shortwave broadcast; amateur radio; aeronautical, marine (coastal waters), military and diplomatic, and scientific research including radio astronomy; and analog cordless phones. For each technology, the simulation model considers the receiver

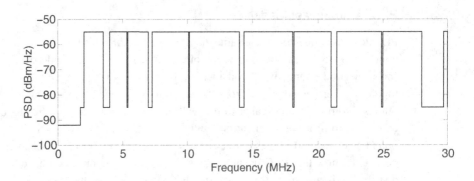

Figure 2.3 IEEE 1901 spectral mask and notches for EMI avoidance. IEEE 1901-2010. Adapted and reprinted with permission from IEEE. © IEEE 2010. All rights reserved.

sensitivity, center frequency (multiple), bandwidth, antenna gains, and noise level. The propagation models consider the near-field effects, which occur within approximately one radian wavelength ($\lambda/2\pi$) of the antenna (cable), the line-of-sight far-field propagation beyond the near field, in which received power decreases in proportion to the square of the distance, and the ground/sky wave emissions that propagate via ground or air, respectively. Based on the simulations and data sets from users and spectrum analysis, Table 2.1 shows that EMI caused by PLC can affect high-frequency technologies unless power control and smart notching are implemented. Power control minimizes the PLC transmission power such that the effective data rate is feasible. It was implemented in PLC devices starting in 2010. Smart notching may detect the presence of victim systems that PLC devices can harm and apply a notch. In June 2010, Ofcom concluded that although power control and smart notching are already part of the PLC devices or product goals of vendors, these features should be formalized to ensure that all vendors comply with EMI requirements. IEEE 1901 specifies the spectral mask up to 30 MHz and dynamic notching for interference reduction between PLC and shortwave radio broadcast [7]. Hence further regulations should enforce the very-high-frequency notches from 30 to 300 MHz.

Ofcom mainly focuses on simulations and does not consider the effect of electrical appliances on EMI caused by PLC. Zarikoff and Malone [10] investigate the effects of electrical appliances on PLC channel transfer function and impedance. They show with a real testbed that PLC devices coexisting with commodity appliances do radiate and that this could affect operations in amateur radio bands. Interestingly, the authors show that the aforementioned standardized frequency notches are not adequate due to non-linear mixing of PLC signals with other appliances attached to the distribution network. Nonlinear appliances modify the PLC signals and create new unwanted frequency components that cause more interference than the original notched PLC signal. Figure 2.4 shows an example of nonlinear signal distortions that lead to IEEE 1901's notches "in-filling." Zarikoff and Malone conduct PLC experiments with a compact fluorescent bulb, a dimmer switch, and a notebook power supply unit connected to the electrical grid. They conclude that the nonlinear effect on PLC signals is not as severe for the compact

Table 2.1 Probability of EMI caused by PLC for certain HF users in 2–30 MHz, according to Ofcom's study [9]

High-frequency (HF) technology	Probability of interference
Aeronautical groundstations	
Power control only	>20%
With power control and notching	<1%
Amateur radio	
Default IARU notches only	>20%
With IARU notches and power control	<1%
Shortwave broadcast listener	
Power control only	>20%
With power control and notching	<1%

Note. Similar results have been found for VHF users in 30–300 MHz. Smart notching is required for negligible probability of interference by PLC. IARU stands for International Amateur Radio Union.

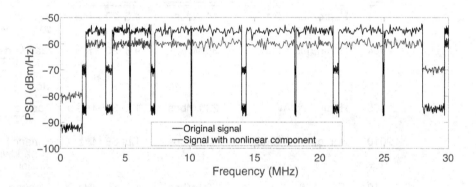

Figure 2.4 Simulated example of infilling notches.

fluorescent bulb and the dimmer switch. The notebook power supply generates non-linear components, and the authors observe that attaching more electrical appliances leads to impedance changes at the IEEE 1901 notches, hence the infilling of the PLC notches. Unwanted emissions of existing PLC devices can be suppressed by using low pass filters such as RF ferrite chokes. Ferrite choke is a passive electric component that suppresses high frequency noise in electronic circuits. Zarikoff and Malone demonstrate that appropriate RF chokes yield a 8–15 dB reduction of PLC emissions and solve this nonlinearity issue in the presence of certain appliances. However, this work has two limitations: First, RF choke installation is not a trivial solution for commodity PLC devices and the end user. Second, the study is limited to three appliances and a single scenario where RF choke appears to achieve the notching. Also, the authors observe that ferrite chokes appear to accentuate the PLC RF emissions in some parts of their test trace. Hence more work is required from PLC vendors to ensure low EMI under non-linear transfer functions for PLC signals. Even though vendors follow the transmission mask requirements of IEEE 1901, careful design is needed to avoid notch infilling due to either nonlinear appliances or high impedance on the PLC channel. This is an open research topic that requires several field trials under different scenarios and power grid environments.

2.3   PLC Specifications and Their Capacity

Table 2.2 presents the most popular PLC specifications that enable 10 Mbps–2 Gbps performance. HomePlug Alliance has released various specifications. The first specification, HomePlug 1.0.1, supports rates up to 14 Mbps. The most recent one, AV2, reaches a 2 Gbps rate; it was released in 2012.

Table 2.2 PLC specifications and standardization in ascending date order

Year	Specification	PHY rate	Bandwidth	Purpose
2001	HomePlug 1.0.1	14 Mbps	4.5–20.7 MHz	"No-new-wires" connectivity
2004	HomePlug Turbo	85 Mbps	4.3–20.9 MHz	Standard-definition video, IPTV, multi-room DVR network, etc.
2005	HomePlug AV	200 Mbps	1.8–30 MHz	Applications such as HDTV and VoIP
2010	HomePlug AV500	500 Mbps	1.8–68 MHz	Enhanced AV rates, by bandwidth extension
2010	IEEE 1901	500 Mbps	1.8–50 MHz	Standardization
2010	HomePlug GreenPHY	10 Mbps	1.8–30 MHz	Automation, comfort, and smart grid
2012	HomePlug AV2	1.5–2 Gbps	1.8–86.1 MHz	4K Ultra HD video and bandwidth-hungry applications

In 2010, IEEE published the 1901 standard [7] for broadband PLC in the band of 1.8–30 MHz, with a possibility of an extension to 50 MHz. Compared to the HomePlug specifications, IEEE 1901 standardizes both in-home and access PLC networks. The relationship between the HomePlug Alliance and the IEEE 1901 standard is similar to that between the Wi-Fi Alliance and the IEEE 802.11 standard [11]. HomePlug certifies PLC products to conform to the 1901 standard and releases enhanced specifications that achieve higher rates that are fully interoperable with previous solutions.

HomePlug AV preceded the IEEE 1901 standard and works on the 1–30 MHz bandwidth. HomePlug AV500 extends this bandwidth to 68 MHz. Finally, AV2 modifies many PHY and MAC layer parameters and also introduces new protocols and functions for higher rates, smaller overhead, and better efficiency.

In addition to solutions for bandwidth-hungry applications, HomePlug released the

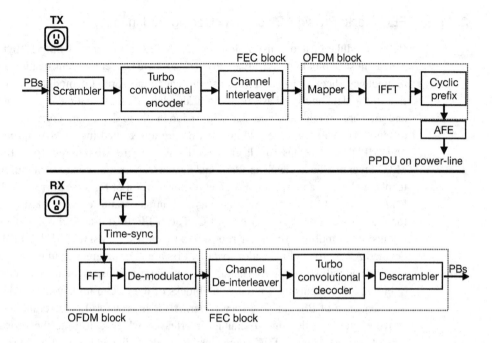

Figure 2.5 PLC transceiver. IEEE 1901-2010. Adapted and reprinted with permission from IEEE. © IEEE 2010. All rights reserved.

GreenPHY specification for home automation, electric vehicle communication, and smart grid. GreenPHY operates at the same frequency band as the other HomePlug specifications, hence it is fully interoperable and can interact with them. Compared to high-rate solutions, GreenPHY typically offers lower rates and lower power consumption.

In the following sections, we review the main features of the PHY layer for the most popular PLC specifications, HomePlug AV and AV2. We explain how the aforementioned PHY rates are achieved.

2.4 PLC OFDM-Based Transceiver

We describe the building blocks of the IEEE 1901 PHY layer for in-home networks. The PHY layer receives data from the 1901 MAC layer and uses fast Fourier transform (FFT) orthogonal frequency-division multiplexing (OFDM) for modulation.[4]

---

[4] The standard also defines wavelet-based modulation, which is not within the scope of this book.

2.4.1   FEC Block: Physical Blocks and Coding Redundancy

Before modulation and transmission, the data in PLC is segmented into 512-byte blocks, called physical blocks (PBs).[5] Each PB has an 8-byte header for identification and cyclic redundancy check (CRC). CRC is used to check the correct demodulation or decryption of the data and is a hash function of the data. PBs are first encrypted for security and then are individually encoded with redundancy, to detect and correct errors in the channel. Typically, a large number of PBs are aggregated into a MAC protocol data unit (MPDU), as we describe later in Section 3.2. Aggregation of multiple PBs reduces MAC-layer overheads, such as additional headers, contention backoffs in multiuser protocols, and interframe spaces. Efficient retransmission of only the corrupted PBs in an MPDU with selective acknowledgments and an automatic-repeat-request (ARQ) protocol addresses impulsive noise in PLC. The IEEE 1901 standard enables PBs to be retransmitted multiple times for robustness (even within a single MPDU) [7], but the detailed ARQ protocol is vendor specific. PBs are the fundamental units of PLC transmission and will be very important in estimating error rates and actual capacity.

Figure 2.5 presents the fundamental blocks of the PLC transceiver. MPDUs are fed as a sequence of PBs into the transceiver. A PB is processed by a scrambler, a turbo convolutional encoder, and a channel interleaver, which constitute the forward error correction (FEC) block. FEC is responsible for detecting and correcting channel errors by converting k data bits into n coded bits that include some redundancy. The ratio k/n is the code rate. The lower the code rate is, the higher the redundancy and robustness of the system are, at the cost of lower effective data throughput.

FEC block starts with the scrambler. The scrambler aims to avoid long sequences of 0s and 1s in the data, as these can yield a high peak-to-average-power ratio that decreases the efficiency of OFDM systems. The scrambler performs exclusive OR operations (XOR) on the data with a predefined pseudo-random polynomial. The receiver descrambles the data using the same polynomial.

FEC then employs the so-called turbo convolutional codes (TCC) that are widely used in many communication systems, such as LTE and WiMAX. TCC are powerful codes that can perform close to the channel capacity and rely on a parallel concatenation of two convolutional codes,[6] with an interleaver between the two codes and an iterative decoder that passes information between the codes [12]. The FEC for PLC was chosen based on simulations and comparison between low-density parity check codes (LDPC) and turbo convolutional coding (TCC). A study over measured power line impulse responses and noise samples showed that TCC codes perform better by 1 dB compared to LDPC for low signal-to-noise ratios (SNRs) [3].

⌁ Further Reading   The PLC TCC encoder consists of two recursive systematic convolutional (RSC) constituent codes and one turbo interleaver, as we see in Figure 2.6. RSC are called recursive because the encoding of current input depends on past encoded

---

[5] In addition, 136-byte and 16-byte block sizes are supported for small packets and acknowledgments.
[6] Concatenated codes like TCC are constructed from two or more simpler codes to achieve high efficiency with reasonable complexity.

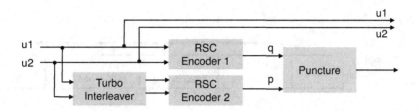

Figure 2.6 The PLC TCC design of concatenated codes and interleaving. IEEE 1901-2010. Adapted and reprinted with permission from IEEE. © IEEE 2010. All rights reserved.

outputs. The TCC outputs are k bits of data (depicted as $u1$, $u2$ in the figure) and m redundant bits (q, p in the figure). The two RSCs are similar and have eight states. RSC encoder 1 operates on data input, whereas RSC encoder 2 operates on the interleaved data input. The turbo interleaver interleaves the PB in dibits, not bits, thus keeping the original bit pairs together. The interleaver length (i.e., the length of the interleaver input and output sequences) depends on the FEC block size (i.e., 16, 136, or 512 bytes). Interleaving distributes the continuous channel errors over different blocks to minimize the probability of decoding error. In addition, the turbo interleaver adds randomness in the data while maintaining its weight (total number of 1s). Puncturing is introduced to reduce the number of parity bits (q, p in the figure) and to achieve high efficiency. The puncturing pattern determines which bits to drop, depends on the FEC rate (1/2, 16/21, and optionally 8/9), and is defined by the IEEE 1901 standard. 📖

TCC is followed by a channel interleaver. The channel interleaver shuffles the data in a predefined manner by the standard, so that channel errors are uniformly spread over the FEC PHY block during decoding. This results in a low error rate under impulsive noise and bursty errors.

The outputs of these three encoders (scrambler, TCC, interleaver) are then fed into an OFDM modulation block, consisting of a mapper, inverse fast Fourier transform (IFFT) processor, preamble, and cyclic-prefix insertion. As illustrated in Figure 2.7, after the FEC block, the MPDU is converted to a physical protocol data unit (PPDU), which is the actual entity transmitted over the power line, by adding a preamble in order to ensure detection by other stations and backward compatibility with previous PLC specifications. The preamble waveform consists of a repeating patten of 7.5 plus-synchronization (SYNCP) symbols followed by 1.5 minus-synchronization (SYNCM) symbols. The SYNCP symbol has a duration of 5.12 μs, and the SYNCM symbol is the same as the SYNCP symbol, apart from its phase, which is shifted by 180°. Each MPDU is preceded by its frame control, called the start-of-frame (SoF) delimiter for data MPDUs. According to the standard, the delimiter is defined as the set of PHY layer fields that precede the payload field in the transmission. The frame control field consists of 128 information bits that are encoded and modulated in one OFDM symbol, which has a duration of 40.96 μs and an effective guard interval of 18.32 μs. The frame control

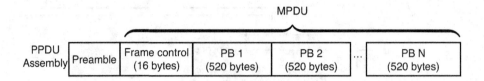

Figure 2.7 PPDU contents. The FEC parity bits are not shown, as they depend on the rate.

signal is transmitted at 0.8 dB higher power than the MPDU's payload and with a 1/2 FEC rate to ensure error-free reception.

2.4.2   OFDM Block

As we saw in Section 2.2.1, the PLC channel suffers from multipath effects due to impedance mismatches caused by electrical appliances attached. The total received signal at a center frequency f_c, $r(t) = \left[\sum_{n=0}^{N(t)} \alpha_n(t)e^{-j2\pi\tau_n(t)}u(t - \tau_n(t))\right]e^{j2\pi f_c t}$ is expressed as a function of the $N(t)$ different multipaths at time t. The path $n = 0$ is the direct nonreflected path, and the unit step function $u(t - \tau_n(t))$ describes the arrival of all subsequent reflected paths with delay $\tau_n(t)$. For any reflected signal n, the multipath channel is described with its amplitude $\alpha_n(t)$ and phase delay $e^{-j2\pi\tau_n(t)}$. It is easy to see that the amplitude and delay of multiple paths can cause interference to subsequent analog symbols, because of delayed arrivals overlapping with the next symbol. The metric that captures the impact of such multipaths and intersymbol interference (ISI) is the delay spread:

$$\bar{\tau}^2 = \frac{\sum_n \alpha_n^2 \tau_n^2}{\sum_n \alpha_n^2}. \tag{2.4}$$

Delay spread can be used to determine the communication symbol duration such that ISI is eliminated. In the frequency domain, the channel can be characterized by the coherence bandwidth B_C, which is inversely proportional to delay spread. If the symbol duration is smaller than the delay spread (or equivalently, if the signal bandwidth is larger than the coherence bandwidth of the channel), then the channel suffers from ISI and frequency-selective fading, where different frequency components of the signal therefore experience different fading. To address such unwanted fading, wireless and power line communications use OFDM, which converts a wideband channel to narrow subchannels (or subcarriers), such that the subcarrier bandwidth is much smaller than the channel coherence bandwidth and ISI is mitigated.

OFDM is used by multiple wideband communication technologies to increase modulation efficiency and address issues caused by multipath channels. It is a frequency division scheme that divides a wideband channel into a multicarrier system, where all subcarriers are orthogonal to each other. Orthogonality ensures that the subcarriers do not interfere with each other and is achieved by sufficient subcarrier spacing. If T is the OFDM symbol duration, then subcarrier spacing typically needs to be at least $\Delta f = 1/T$.

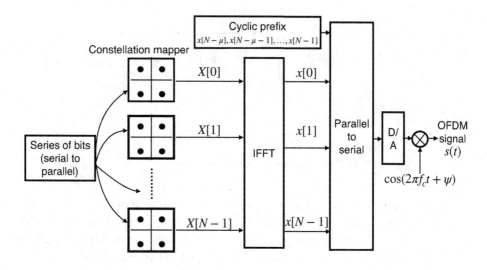

Figure 2.8 OFDM signal generation.

This would yield $N = B/\Delta f$ subcarriers if the PLC bandwidth were B. Also, the subcarrier spacing needs to ensure that each subcarrier faces an approximately frequency-flat channel, hence $B/N < B_C$. An OFDM waveform is constructed by an inverse Fourier transform of subcarrier coefficients $X[k]$ and yields the finite duration signal

$$x(t) = \frac{1}{N} \sum_{k=0}^{N-1} X[k] e^{j2\pi \cdot k \cdot \Delta f \cdot t}, 0 \leq t \leq T, \tag{2.5}$$

where N is the number of samples in inverse Fourier transform. Even with Δf sufficient subcarrier spacing, orthogonality between subcarriers can degrade due to additional intersymbol interference. Hence additional techniques are applied to avoid ISI, as we explain below. The finite duration signal $x(t)$ then passes through digital to analog conversion and is modulated up to the center frequency f_c, to produce the transmission signal

$$s(t) = Re\{e^{j2\pi f_c t} \sum_{k=0}^{N-1} X[k] e^{j2\pi \cdot k \cdot \Delta f \cdot t}\}. \tag{2.6}$$

Figure 2.8 shows the generation and typical transmitter design of OFDM signals. We next give details on the practical implementation and the cyclic prefix inserted in addition to data symbols in order to avoid ISI.

Both PLC and Wi-Fi rely on FFT for practical modulation and demodulation. FFT enables cheap and practical OFDM signal generation. Before FFT implementations, signal generation was the main barrier for OFDM schemes. As we show in Figure 2.8, data is first converted from serial to parallel N subcarriers. At the transmitter, the signal is generated in the frequency domain with the mapper that chooses the appropriate phase and amplitude for each subcarrier (frequency), based on the digital data per subcarrier.

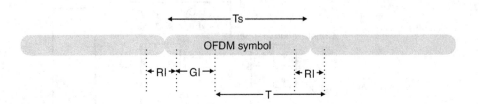

Figure 2.9 PLC OFDM symbol structure and timing.

Each OFDM subcarrier i corresponds to one element of the discrete Fourier spectrum $X[i]$. The amplitudes and phases of the subcarriers depend on the transmitted data (see the next section, on modulation, for more details). After mapping digital data to subcarrier frequency components, IFFT is performed on each subcarrier data to convert $X[i]$s to time domain $x[i]$s, and a cyclic prefix is added before the parallel data are converted to serial again. This time-domain data is converted to an analog signal using a digital-to-analog converter (D/A).

✎ Further Reading An IFFT generates the signal from mapped data sequence $X[k]$, $0 \le k \le N - 1$, of N subcarriers:

$$IFFT\{X\} = x[n] = \frac{1}{N} \sum_{k=0}^{N-1} X[k] e^{2\pi \cdot j \frac{n \cdot k}{N}}, \tag{2.7}$$

where $x[n]$ are the time-domain samples and $X[k]$ the frequency-domain ones. The output of a linear time-invariant system, such as a channel, is equal to the convolution of the input signal with the channel impulse response. Then, at the receiver, we have the signal (ignoring noise)

$$y[n] = x[n] * h[n], \tag{2.8}$$

which is the convolution of the input signal with the channel impulse response $h[n]$. The convolution in the time domain is equivalent to multiplication of the FFT transforms in the frequency domain. In the FFT domain, the input $x[n]$ can be retrieved simply by $X[i] = Y[i]/H[i]$, where $Y[i]$ is the FFT of the channel output at the receiver and $H[i]$ is the FFT of the channel impulse response. The equivalent convolution in FFT is called circular convolution. The circular convolution is defined as

$$y[n] = x[n] \circledast h[n] = \sum_k h[k]x[n - k]_N, \tag{2.9}$$

where $[n - k]_N = (n - k) \mod N$ is a periodic repetition of $x[n - k]$ with period N. However, a linear time-invariant channel produces linear convolution and not circular, and FFT relationship $X[i] = Y[i]/H[i]$ cannot be directly used at the receiver. Hence the OFDM scheme needs to ensure that circular convolution is performed by the channel. A multipath channel of length μ can be described by the transfer function $\mathbf{h} = [h_0 \, h_1 \, \dots \, h_{\mu-2} \, h_{\mu-1}]$. To achieve circular convolution, most OFDM schemes repeat a sequence of μ mapped data at the beginning of the OFDM symbol. This sequence of data $x[i]$, $N - \mu \leq i \leq N - 1$, of N is called the cyclic prefix and is appended such that the final transmitted sequence is $\tilde{\mathbf{x}} = x[N - \mu], x[N - \mu - 1], \ldots, x[N - 2], x[N - 1],$ $x[0], x[1], \ldots, x[N - 2], x[N - 1]$, as we see in Figure 2.8. The cyclic prefix has two advantages: (1) it allows the linear convolution of a frequency-selective multipath channel to be modeled as circular convolution and leads to simple frequency domain processing, such as channel estimation and equalization, and (2) it provides a guard interval to minimize ISI from the previously transmitted symbol. Figure 2.9 shows OFDM symbol timing and the guard interval (GI) of the PLC symbol. Let T_g be the duration of the cyclic prefix and T the effective symbol duration. Then, $T_s = T + T_g$ in the figure. Given that N samples correspond to a whole symbol T, we have $\mu = T_g/T$. The cyclic prefix insertion introduces overhead and reduces the effective data rate by $T/(T + T_g)$, which creates a trade-off between overhead and ISI. After adding the cyclic prefix, the channel output becomes

$$y[n] = \tilde{x}[n] * h[n] = \sum_{k=0}^{\mu-1} h[k]\tilde{x}[n - k] = \sum_{k=0}^{\mu-1} h[k]x[n - k]_N = x[n] \circledast h[n]. \quad (2.10)$$

After the insertion of the cyclic prefix, the channel estimation and equalization can be easily achieved, and the receiver can retrieve $\tilde{x}[n]$ by $X[i] = Y[i]/H[i]$.

OFDM results in a sum of weighted sinc functions ($\text{sinc}(x) = \sin(x)/x$) in the frequency domain, as the Fourier transform of time-duration pulse function $u(t) - u(t - T)$[7] is $\text{sinc}(\pi(f - k\Delta f)T)$ and subcarrier k. Sharp edges of the OFDM symbols in the time domain produce slow decaying sinc functions in the frequency domain, which causes high out-of-band emission in the OFDM spectrum. The solution to out-of-band emission is to use symbol windowing (special filters) to the front and back of an OFDM symbol. In addition to the guard interval in Figure 2.9, we show the roll-off interval (RI) of the PLC OFDM scheme. OFDM symbols are smoothed during RI (usually using raised cosine functions). PLC is using a windowed OFDM scheme to avoid adjacent channel interference. 📖

The PLC OFDM time domain signal, based on a 100-MHz sampling clock, is generated as follows. For 1901 frame control and payload symbols, a set of data points from the mapping block is modulated onto the subcarrier waveforms using a 4096-point IFFT, resulting in 4096 time samples (IFFT interval). A fixed number of samples from the end of the IFFT are taken and inserted as a cyclic prefix at the front of the IFFT interval to create an extended OFDM symbol. The 4096-point IFFT generates 2048 positive-frequency subcarriers (in 0–50 MHz bandwidth), with the subcarrier frequency of subcarrier k equal to $F_k = k \cdot 100 \cdot 10^6/4096$ Hz. Note that some subcarriers are not used due to EMI (e.g. subcarriers $0 \leq k \leq 85$ are OFF). A given subcarrier k is turned off if it violates the spectral mask shown in Figure 2.3 or if the device does not support bandwidths up to 50 MHz (e.g., HomePlug AV supports frequencies only up to 30 MHz – in total, 1024 subcarriers can be used for HomePlug AV, as discussed in

---

[7] The difference of the two step functions represents one symbol.

the next section). The PLC OFDM symbol duration is 40.96 μs, and, to overcome the delay spread of the channel, there is a variable guard interval that can take the values 5.56, 7.56, or 47.12 μs. The OFDM block forwards the symbols to the analog front end (AFE) module that couples the signal to PLC cables.

2.4.3   The PLC Receiver

The PLC receiver is equipped with an AFE with automatic gain control (AGC) and time synchronization that separates frame control and payload data. Data is recovered by processing the received stream through a 4096-point FFT block, a demodulator, and a de-interleaver followed by a turbo convolutional decoder and a descrambler to recover the originally transmitted data. The turbo decoder works iteratively by combining the outputs of the two RSC decoders and a turbo interleaver. The two RSCs iteratively take turns, each decoding one of the underlying convolutional codes. The soft decision output (log-likelihood ratios or probabilities for each information bit) of the first decoder is used as a prior distribution for the other, improving the information bit estimates of the second decoder, and vice versa in the next iteration. This iterative decoding process continues until a terminating condition is met, which is not defined by the standard.

Depending on the channel conditions and the robustness requirements, different modulation schemes and FEC rates can be selected for each link. The receiver of the data estimates the optimal PHY parameters with a process that we discuss in Section 2.7. For example, harsh channel conditions or control messages that require reliability lead to transmissions at robust modulation (see next section) and 1/2 FEC rate, which adds 100 percent redundancy for every bit transmitted. Modulation schemes vary across both frequency and time domains. In the following section, we focus on modulation and coding selection and adaptation, the key parameters of the PHY layer.

2.5   Modulation and Coding Schemes

To overcome the multipath nature of PLC, the PHY layer of HomePlug AV (IEEE 1901) is based on an FFT-OFDM scheme with 1024 subcarriers in the 1.8–30 MHz frequency band. The whole band is used for every frame transmission, with only a few subcarriers notched out to avoid causing interference to other existing transmissions, such as those on amateur radio bands (see Section 2.2.4). The modulation of FFT-OFDM is based on digital quadrature amplitude modulation (QAM) schemes. In QAM, bits are modulated using the amplitudes of two carrier waves, whose phases differ by 90°. QAM can carry a variable number of bits per symbol, with a trade-off between robustness and goodput. If we transfer x bits per QAM symbol, then since bits can have a 0 or 1 value, there are 2^x possible bit sequences per QAM symbol. Each of the bit sequences is mapped to a unique amplitude-phase pair of the analog carrier. This mapping is called QAM constellation. Increasing the data rates and the number of bits per QAM symbol x results in a higher-order constellation where the distance between the mapped amplitude-phase pairs decreases. Since these pairs are closer, the demodulator at the receiver can distinguish adjacent pairs with higher uncertainty under noise and interference; this results in a higher bit error rate. Higher-order QAM is faster but less robust than lower-order QAM, if the mean constellation energy is preserved. Figure 2.10 presents this trade-off using `Matlab` modulation function `qammod`. 30 000 bits are modulated into 7,500 and 5,000 symbols for 16-QAM and 64-QAM, respectively. 64-QAM modulation results in fewer symbols and less transmission time, hence higher goodput. However, the plot suggests that when the noise is the same (left with 16-QAM and center with 64-QAM), the error rate is higher with 64-QAM, leading to demodulation ambiguities. Using `qamdemod` and `biterr` functions, we indeed compute that 64-QAM has unacceptable bit errors for this channel: $3.84 \cdot 10^{-2}$ versus $2.40 \cdot 10^{-3}$ for 16-QAM. Similarly to this example, a PLC receiver adapts the QAM modulation order both in frequency and time domains, based on the error rate.

Channel varies with frequency, as PLC is frequency selective due to multipath effects (see Section 2.2.1). Because of the high frequency selectivity of the channel that yields variable attenuation across frequencies, each PLC OFDM subcarrier can employ a different modulation scheme out of BPSK, QPSK, and 8/16/64/256/1024-QAM. BPSK and QPSK are robust schemes since they carry only 1 and 2 bits per OFDM symbol, respectively. QPSK is used for the initial channel estimation and communication between two stations but also for all broadcast and multicast transmissions. There are three robust modes of communication, called ROBO modes in IEEE 1901: standard, mini, and high-speed ROBO. All use QPSK but with different interleaving techniques, that is, number of data copies for robustness and PB size. The ROBO interleavers create redundancy by reading the output bits multiple times. To ensure adequate frequency separation between the multiple copies of each bit, portions of the initial interleaved output are inserted between successive readouts of the interleaver output. The receiver uses the extra repetitions to detect and correct errors at the receiver. Table 2.3 summarizes the parameters for the ROBO modes.

Table 2.3 IEEE 1901 ROBO modes

ROBO mode	Interleaving copies	Guard interval	PB size	PHY rate
Standard	4	5.56 µs	520 bytes	4.9226 Mbps
Mini	5	7.56 µs	136 bytes	3.7716 Mbps
High-speed	2	5.56 µs	520 bytes	9.8452 Mbps

QAM modulations yield high throughput under good channel conditions, delivering x bits per symbol in any 2^x-QAM scheme. Figure 2.11 shows an example of the modulation selected by two different PLC links, as a function of the subcarrier frequency (or subcarrier index): we plot the bit per symbol averaged over 2 hours for the two links. We notice that indeed, PLC is highly frequency selective, since the modulation schemes simultaneously used by different subcarriers can range from the highest (1024-QAM or 10 bits per symbol) to the lowest (QPSK or 2 bits per symbol). In contrast to PLC,

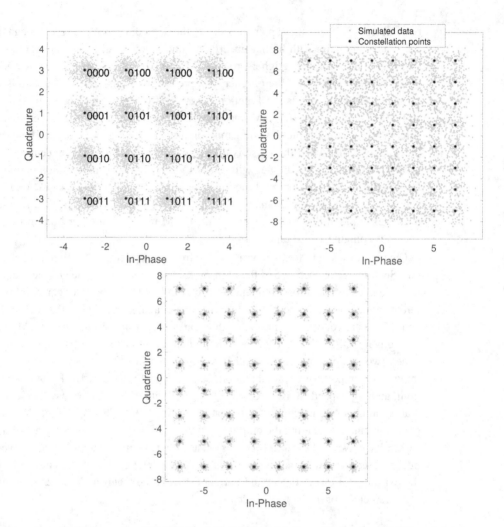

Figure 2.10 Example of 16-QAM and 64-QAM modulations. We simulate 30 000 modulated bits using the two different schemes. The plot shows the constellation map and the received simulated data in a channel of Gaussian noise. For two upper simulations, the ratio of bit energy to noise power spectral density is set to 10 dB. 64-QAM yields high error under 10 dB ratio. In the lower plot, the ratio of bit energy to noise power spectral density is set to 20 dB. 64-QAM performs best under good channel conditions.

Wi-Fi OFDM schemes employ the same modulation across all subcarriers, yielding a much simpler PHY/MAC design than PLC.

In addition to an adaptive modulation scheme, PLC introduces parity bits to reduce random errors in the channel at the receiver demodulator and decoder. As explained in Section 2.4, the FEC code used by HomePlug and IEEE 1901 is TCC. With HomePlug AV, the available rates for FEC are 1/2 and 16/21. For any rate k/n, FEC introduces $(n-k)$ parity bits for k bits of data. Hence the effective PLC PHY-layer throughput is the PHY rates achieved by the OFDM modulation scheme times k/n. TCC is succeeded by

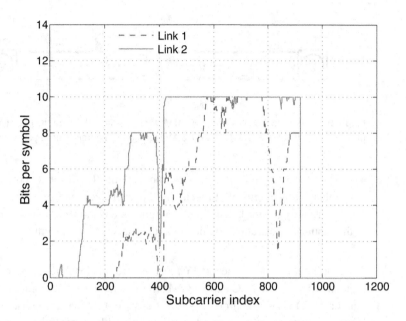

Figure 2.11 Example of PLC modulation. Bits per symbol averaged over two hours as a function of PLC subcarrier for two links of our testbed. Subcarrier index is reported by the PLC device, after removing the notched frequencies (not presented in this plot).

channel interleaving for impulsive noise, prevalent in PLC channels. The combination of OFDM, guard interval with cyclic prefix, TCC, and interleaving enables robust and high-rate communications over a challenging channel.

To achieve robust, low-rate communication, HomePlug GreenPHY uses the same frequency band as AV, but the only available modulation is the QPSK scheme and FEC is the 1/2 rate, using only the ROBO modes in Table 2.3. Owing to these robust schemes, GreenPHY achieves throughput only up to 10 Mbps with an extremely low error rate.

2.6 HomePlug AV2: New Key Features and MIMO Transmission

PLC modulation and coding techniques are continuously evolving. The most recent specification, AV2, introduces additional features to support bandwidth-hungry applications and improved data rates. Compared to AV, it incorporates the following:

- extension of the frequency spectrum to 1.8–86 MHz (vs. 1.8–30 MHz for HomePlug AV) and of FFT points to 8192 (vs. 4096 for HomePlug AV), to accommodate the additional bandwidth and 3455 subcarriers (vs. 917 for HomePlug AV);
- 4096-QAM modulation scheme, 16/18 FEC rate, and 0 μs guard intervals;
- lower MAC overhead via shorter delimiters[8] and delayed acknowledgments;
- support for MIMO (multiple-input and multiple-output) with beamforming;

[8] Delimiters are frame preambles with PHY/MAC information. We describe them in detail in Chapters 3 and 4.

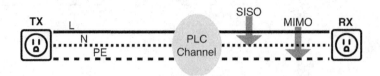

Figure 2.12 Electrical cables used for MIMO (HomePlug AV2) are the line (L), neutral (N), or protective earth (PE). SISO (HomePlug AV) uses only the line-neutral pair of wires.

- support for 32-byte PBs, which can be used in PHY-level acknowledgments – they enable acknowledgment of much larger packet sizes; and
- EMI-friendly power boost, which optimizes the transmission power by monitoring the input-port reflection coefficient at the transmitting modem.

The most significant AV2 new PHY-layer feature is MIMO transmission, which can theoretically double the PLC capacity. MIMO is widely used in communications systems (such as Wi-Fi, LTE, and other wireless systems) to increase capacity and to provide signal diversity gains without additional bandwidth or transmission power. HomePlug AV is a single-input single-output antenna system (SISO) that uses the line and neutral electrical wires. Whereas HomePlug AV uses the line-neutral cable pair for SISO, AV2 employs any two pairs of wires formed by the line (L), neutral (N), or protective-earth (PE) wires (i.e., L-N, L-PE, or N-PE) as two MIMO antennas. The antenna can be any voltage difference between L, N, and PE wires. A comparison between SISO and MIMO for PLC is given in Figure 2.12. A receiver plugged into power outlets that are connected to the three wires L, N, and PE can employ up to three differential-mode receiver ports and one common-mode port. The common-mode signal is the voltage difference between the sum of the three wires and the ground. PLC MIMO requires that a third wire PE be installed at the building, which is not the case for a few old electrical installations. Schwager presents a detailed survey of AV2 features and MIMO [13].

MIMO allows for significantly improved peak data rates and performance by exploiting the multiple paths over different antennas. PLC antennas (cables) are located in parallel within the walls, experiencing a similar multipath propagation. The medium becomes broadcast and MIMO superposition is feasible because of coupling between the wires. HomePlug AV2 MIMO uses two independent transmitters and up to four receivers, with beamforming required to maximize the performance on the independent streams. In practice today, two receivers are used, leading to 2×2 MIMO. The potential MIMO signal paths for PLC are shown in Figure 2.13. Because of the crosstalk between PLC wires, the signals from any TX$_i$ port can be received by all receiver ports RX$_i$, $1 \leq i \leq 4$. The three transmitter and receiver ports are the voltage differences between the wire pairs L-N, L-PE, or N-PE. In addition, the receiver can use the common mode between L, N, PE, and the ground. The transceiver of AV2 is modified compared to the one in Figure 2.5 to add a MIMO analog front end and the appropriate FFT points (up to 8192).

HomePlug AV2 PLC transceivers also aim at efficient notching and EMI-friendly

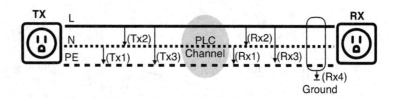

Figure 2.13 PLC MIMO potential transmitter ports (TX$_i$, $1 \leq i \leq 3$) and receiver ports (RX$_i$, $1 \leq i \leq 4$). The individual signals at the receiver are L-N, L-PE, N-PE, and a common mode signal that is the voltage difference between the sum of the three wires and the ground. Reproduced from *An Overview of the HomePlug AV2 Technology*, Hindawi Publishing Corporation. © 2013 Larry Yonge et al.

power boost [13]. PSD limits of HomePlug AV2 and IEEE 1901 are based on representative statistics of the impedance match at the interface between the device port and the power line network. However, the input impedance of the power lines is frequency selective and varies in different environments. As a result, a portion of the transmitted power is lost within the transmitter. The input-port reflection coefficient, or input return loss, is characterized by the so-called parameter S_{11}, and a portion of lost transmission power is $20 \times \log_{10}(|S_{11}|)$ (dB). When the impedance mismatch (hence also S_{11}) is large, only a small fraction of input power is effectively transferred to the power lines, and EMI induced by the PLC modem is reduced to potentially lower than the values recommended by EMI regulations. To compensate for the frequency-selective impedance mismatch at the interface between the device port and the power line network, HomePlug AV2 devices adapt their transmission mask upon measurement of S_{11} at the transmitter. Signal power is increased by an impedance mismatch compensation (IMC) factor, leading to a more effective power transmission of PLC data. An increase of PLC transmission power leads to stronger radiated EMI. Hence the HomePlug AV2 design of the IMC ensures that the resulting EMI continuously satisfies the regulation limits. IMC limits the transmission power increase to a maximum value:

$$\text{IMC}(k) = \min\left(\max\left(10\log_{10}\frac{1}{1 - |S_{11}(k)|^2} - M(k), 0\right), \text{IMC}_{\max}\right), \qquad (2.11)$$

where k is any subcarrier, $S_{11}(k)$ is the estimate of S_{11} for this subcarrier, and $M(k)$ is a margin for accounting for uncertainties in the real-time measurement of S_{11}. The maximum IMC values and modeling were determined statistically by a series of measurements in six European countries [13]. This is the most recent and comprehensive study of PLC EMI, with both indoor and outdoor tests [14]. An open research topic would be the repetition of EMI tests with commercial devices, which could implement different IMC configurations.

📖 **Further Reading** We will now analyze the benefits of MIMO mathematically, with the HomePlug AV2 2×4 antenna configuration. Figure 2.14 presents all possible paths

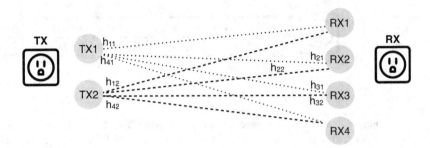

Figure 2.14 Example of HomePlug AV2 2×4 MIMO channel. The transmit streams go through a channel matrix **H**, which consists of all eight paths between the two transmit antennas and four receive antennas.

of a signal stream, which form a 2×4 channel matrix:

$$\mathbf{H} = \begin{bmatrix} h_{11} & h_{21} \\ h_{21} & h_{22} \\ h_{31} & h_{32} \\ h_{41} & h_{42} \end{bmatrix}. \tag{2.12}$$

The channel matrix represents the channel gain from any transmit/receive antenna pair. For any input vector **x**, the received output can be modeled as $\mathbf{y} = \mathbf{H}\mathbf{x} + \mathbf{z}$, where **z** is the channel noise. By knowing the channel matrix **H**, the transmitter can optimize the transmission depending on channel gains h_{ij}, a technique known as precoding or beamforming. Depending on the MIMO design, one or two independent streams can be transmitted over the two transmit ports TX_1 and TX_2. Beamforming requires knowledge about the channel state information at the transmitter to perform optimal precoding. The full channel state information is available at the receiver, hence feedback is required for precoding. As we note in the next section, HomePlug relies on feedback from the receiver for adaptive modulation and coding. Hence HomePlug AV2 can leverage similar feedback for MIMO adaptive precoding. 🔍

We now explain how precoding works in MIMO channels. The channel matrix can be factorized using singular value decomposition (SVD):

$$\mathbf{H} = \mathbf{U}\mathbf{\Sigma}\mathbf{V}^*, \tag{2.13}$$

where **U** is a 4×4 unitary matrix (i.e., $\mathbf{U}^* = \mathbf{U}^{-1}$), $\mathbf{V}^*$ is a conjugate transpose of matrix **V**, $\mathbf{\Sigma}$ is a 4×2 rectangular diagonal matrix with nonnegative numbers on the diagonal, and **V** is a 2×2 unitary matrix. SVD is a widespread technique to factorize matrices in signal processing and statistics. In MIMO communications, SVD converts a MIMO channel into parallel SISO channels through transmit and receive beamforming, if the channel matrix is available in both terminals. To achieve this, the transmitter premultiplies its signal with the matrix **V**, and the receiver multiplies the channel output by $\mathbf{U}^*$, forming the effective channel $\widetilde{\mathbf{y}} = \mathbf{\Sigma}\widetilde{\mathbf{x}} + \widetilde{\mathbf{z}}$, where $\widetilde{\mathbf{y}} = \mathbf{U}^*y$, $\widetilde{\mathbf{x}} = \mathbf{V}^*x$, and $\widetilde{\mathbf{z}} = \mathbf{U}^*z\mathbf{V}$. Note that $\mathbf{\Sigma}$ is a 4×2 rectangular diagonal matrix with diagonal values σ_i, hence the two effective parallel channels are given by $\widetilde{y}_i = \sigma_i\widetilde{x}_i + \widetilde{z}_i$, with $0 \leq i \leq 2$.[9] In the case of white Gaussian noise $\mathbf{z}$, owing to spherical symmetry, the effective noise $\widetilde{\mathbf{z}}$ remains white Gaussian after the unitary transformations $\mathbf{U}^*$ and $\mathbf{V}$.

For precoding, the matrix $\mathbf{V}$ has to be fed back to the transmitter by the receiver, and precoding is performed between the mapper and the IFFT of the transceiver in Figure 2.5, where each MIMO stream would have a different parallel OFDM block. The beamforming matrix is quantized efficiently at the receiver to reduce the amount of feedback required for $\mathbf{V}$. This quantization is chosen such that the loss of the SNR after detection is within 0.2 dB compared to optimum beamforming without quantization.

2.7   Channel Estimation and Adaptation

Because each OFDM subcarrier can employ a different modulation scheme, PLC stations exchange messages containing information about the modulation scheme that each subcarrier uses (i.e., the number of bits per symbol), the FEC rate, and the guard interval length. The set of all these parameters is called the tone map. It is determined during the channel-estimation process for every link in the PLC network. A station that has data to send and determines that it has no valid tone maps for communication starts the initial channel estimation to the destination of the data depicted in Figure 2.15. The source initially sends "sound" frames to the destination by using a default, ROBO modulation scheme that uses QPSK for all subcarriers. The sound frames have a sound reason code set to indicate sounding for initial channel estimation. The destination of the data responds with a sound MPDU, with the sound ACK flag (SAF) bit in the frame control set to 0b1 and with no payload. This sound frame exchange continues until the destination determines that it has sufficient data for the tone map estimation, in which case it transmits with a sound MPDU with SAF = 0b1 and sound complete flag SCF = 0b1. The destination should receive no more than 20 ms duration of ROBO sound frames. After estimating the tone map, the receiver sends the tone map and its unique identification, called the tone map index, back to the source with a CM_CHAN_EST.indication management message. The source stores the tone map associated with the tone map index and starts using it immediately. Each transmitted MPDU is tagged with the index of the tone map that has been used in order to inform the receiver about the demodulation procedure (i.e., which tone map to use for demodulation). The receiver can choose up to $L + 1$ tone maps, for some integer L: L tone maps for different subintervals of the AC line cycle, called slots, and one default tone map. PLC uses multiple tone maps for the different slots of the AC line cycle to mitigate the periodic impulsive noise that is synchronous to the mains. Devices are synced using beacons to align tone maps with AC synchronous noise.

Katar et al. [15] investigate the optimal duration of the tone map slots, and they

---

[9] For a general architecture with n_t transmit antennas and n_r receive antennas, the channel vector consists of $n_{min} = \min(n_t, n_r)$ singular values.

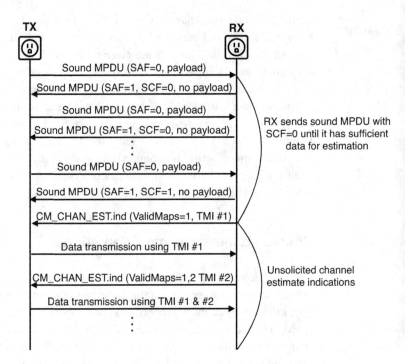

conclude that 1–2 ms duration can result in significant capacity gains. Our experimental results uncover that, in practice, there are $L = 6$ tone maps chosen and used in the first half of the AC line cycle and then repeated periodically (see also Figure 2.17 [top] where we have six periodic tone maps, T_i, $0 \le i \le 6$). For 50-Hz mains, the duration of one-half of the AC line cycle is 10 ms, which means that, indeed, the tone map slot has a duration of $10/6 = 1.67$ ms. IEEE 1901 also defines the use of a maximum seven tone maps for the AC cycle, with six tone maps for the AC intervals and a default tone map. The AC cycle intervals are disjoint. The default tone map can be used anywhere in the AC period.

Tone maps are updated dynamically, either when they expire (after 30 s) or when the receiver of the data indicates that tone maps are invalid. This can happen, for example, when the PB error rate or SNR exceeds a threshold. When channel conditions change, the receiver invalidates a tone map either by transmitting a CM_CHAN_EST.indication frame with a valid tone map list that does not include the invalidated tone map or with a bit flag in the SACKI field of an acknowledgment. If the receiver invalidates the default tone map or the whole tone map list, then channel estimation is performed for the whole AC cycle. Otherwise, it can be performed for a specific AC interval. Figure 2.16 shows the frame exchange in dynamic channel estimation. The exact channel-estimation

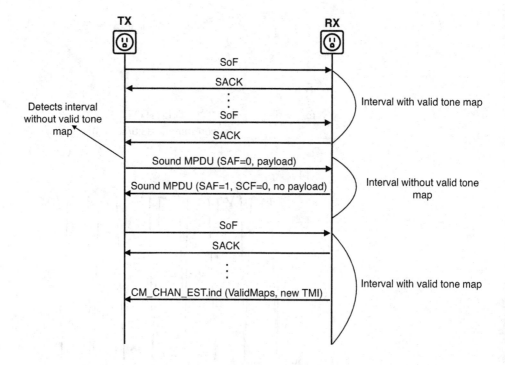

Figure 2.16 Dynamic channel estimation procedure for PLC. IEEE 1901-2010. Adapted and reprinted with permission from IEEE. © IEEE 2010. All rights reserved.

algorithm at the receiver based on PB error rate and other metrics is not defined by the IEEE 1901 standard and is vendor specific.

2.7.1 Bit-Loading Estimate

The bit-loading estimate (BLE) is an estimation of the total number of bits that can be carried on the channel per microsecond. BLE is defined by the IEEE 1901 standard as follows [7]. Let T_{sym} be the OFDM symbol length in μs (including the guard interval), R be the FEC code rate, and PB_{err} be the PB error rate (estimated based on the expected PB error rate on the link when a new tone map is generated, which remains fixed until the tone map becomes invalidated by a newer tone map). Let also B represent the sum of the number of bits per symbol over all subcarriers, determined by the tone map. Then, BLE is given by

$$BLE = \frac{B \cdot R \cdot (1 - PB_{err})}{T_{sym}}. \tag{2.14}$$

BLE takes into account the overhead due to FEC but does not include any MAC-layer overhead, such as frame-control duration or interframe spaces. It is an estimation of the

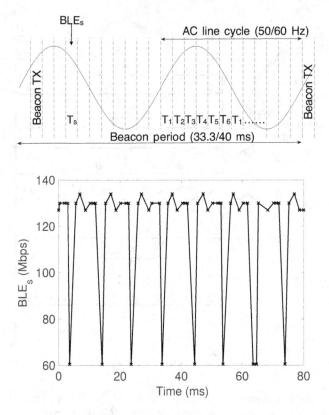

Figure 2.17 (top) Periodicity of tone maps: example of tone map slots. For any slot s, there exists a tone map with ID T_s yielding the rate BLE_s. (bottom) Measurement of BLE on a real link of our testbed.

PHY-layer capacity, as we observe in Chapter 5. The bit loading is unique per pair of stations, and it can vary within the mains cycle, as we explained above.

As explained in Section 2.4.1, each MPDU is preceded by its frame control, called the start-of-frame (SoF) delimiter. BLE can be retrieved from SoF. The SoF contains information for both PHY and MAC layers, such as BLE or MPDU duration.

Figure 2.17 presents an example of BLE periodic design and a snapshot of BLE versus time for a link of our testbed. As we observe, BLE for a single link can vary from 60 to 130 Mbps, because channel quality varies within the mains cycle. Section 2.2.2 explains that this variability is due to periodicity of noise (periodic noise synchronous to the mains).

2.8   IEEE 1901 PHY Parameters

We now summarize the main PLC PHY-layer parameters. Table 2.4 presents the IEEE 1901 PHY-related PLC parameters and Table 2.5 the PHY parameters for HomePlug

AV2. As we saw in Section 2.4.2, IEEE 1901 and HomePlug AV use an IFFT of 4096 points, which generates up to 2048 OFDM subcarriers. However, not all subcarriers are used for transmission due to EMI regulations, as we notice in Section 2.2.4. The IEEE 1901 FFT PHY uses bandwidth from 1.8 MHz to 50 MHz, but support for subcarriers above 30 MHz is optional. The maximum supported subcarriers are 1024 in frequency range 1.8–30 MHz and 2048 over the maximum specified bandwidth (1.8–50 MHz). Of the 1024 mandatory subcarriers below 30 MHz, 917 are used for modulation for the default broadcast and extended tone maps. The optional 50 MHz maximum bandwidth configuration uses up to 1974 subcarriers out of 2048 due to EMI spectral mask. The rest of the subcarriers are masked out to avoid interference on AM and radio amateur bands. The subcarrier spacing is 24.414 kHz.

HomePlug AV uses 1.8–30 MHz, up to 1024 QAM modulation schemes, and 1/2 and 16/21 FEC rates. HomePlug AV2 extends bandwidth to 86 MHz (which is beyond the maximum 50 MHz bandwidth defined by IEEE 1901 standard), modulation up to 4096 QAM (which is an optional feature of IEEE 1901), and it additionally uses 16/18 FEC rate (another optional feature of IEEE 1901). By using these extensions in addition to 2×2 MIMO, HomePlug AV2 supports up to 1.5 Gbps rates.

All PLC devices are required to support up to seven tone maps: six for the AC cycle slots and one default tone map.

Finally, IEEE 1901 defines various guard intervals (GI) to mitigate intersymbol interference and multipath effects. Depending on the data rate, different GI can be used, with higher data rates employing shorter GI (e.g., HomePlug AV2).

Table 2.4 IEEE 1901 PHY parameters

Parameter	Value
Frequency band	1.8–30 MHz (mandatory) and 30–50 MHz (optional)
IFFT/FFT points	4096
Maximum number of subcarriers	1024 (1.8–30 MHz) or 2048 (1.8–50 MHz)
Subcarrier spacing	24.414 kHz
Number of used subcarriers due to EMI	917 (1.8–30 MHz) or 1974 (1.8–50 MHz)
Modulation schemes	BPSK, QPSK, 8/16/64/256/1024/4096 (optional) QAM
Symbol duration	40.96 µs
Roll-off interval (RI)	4.96 µs
Guard interval (GI)	5.56 µs, 7.56 µs, 47.12 µs
Cyclic prefix	RI + GI
Maximum number of tone maps	7
FEC rates	1/2, 16/21, 16/18 (optional)

2.9   IEEE 1901 versus G.hn

We provide a high-level comparison between the IEEE and G.hn standards related to PLC in Table 2.6. First, G.hn 9960 provides PHY-layer specification for coaxial and phone lines in addition to power lines, using a different subcarrier spacing for every medium [16]. Contrarily, IEEE 1901 only specifies communications over power lines.

Table 2.5 HomePlug AV 2 parameters

Parameter	Value
Frequency band	1.8–86 MHz
IFFT/FFT points	8192
Maximum number of subcarriers	4096
Number of used subcarriers due to EMI	3455
Modulation schemes	BPSK, QPSK, 8/16/64/256/1024/4096 QAM
Symbol duration	40.96 µs
RI	4.96 µs
GI	5.56 µs, 7.56 µs, 47.12 µs
Cyclic prefix	RI + GI
Maximum number of tone maps	7
FEC rates	1/2, 16/21, 16/18

IEEE 1901 provides both home and access environments and wavelet and OFDM modulation schemes. However, G.hn only focuses on home networking (not access) and only on FFT-OFDM modulation (not wavelet-OFDM). FEC is also different between the two technologies, with G.hn using LDPC codes and IEEE 1901 using TCC for FFT-OFDM covered in this book. Wavelet-OFDM uses concatenated Reed–Solomon (RS) and convolutional codes in IEEE 1901. MIMO transmission is not defined in IEEE 1901 but is specified later with HomePlug AV2 specification. The ITU-T G.9963 standard defines MIMO transmissions for G.hn networks [17].

These differences on the PHY layer mean that the IEEE and G.hn technologies cannot decode each other's frames. To address this issue, IEEE 1901 and ITU-T G.cx (or G.9972) standardize the same coexistence ISP (Inter System Protocol) between the two technologies [18]. G.9972 and ISP provision two mechanisms for coexistence of different systems over the power line.[10] The first one uses frequency-division multiplexing (FDM), in which the bandwidth is divided into two parts, with frequencies below 10 or 14 MHz (the specific value can be selected by an access network) and frequencies above. The second one uses time-division multiplexing (TDM), in which the available airtime is divided equally between two technologies. Frequency or time allocation is achieved by coexistence signaling, which uses periodically repeating ISP windows that convey information on coexisting system presence, resource requirements, and resynchronization requests. Each coexisting system category is allocated a particular ISP window in a round robin fashion. The ISP signal consists of sixteen consecutive OFDM symbols. Each OFDM symbol is formed using a 512-point IFFT, with each subcarrier having a phase shift defined in G.9972. The ISP time domain signal is then multiplied by a window function to reduce the out-of-band emissions and to be compliant with the transmit spectrum mask. The window function is vendor proprietary. The ISP window contains two ISP fields, which determine the presence, mode, or request for frequency or time allocation of a particular network.

---

[10] Different systems can be G.hn, IEEE 1901 Internet access or IEEE 1901 in-home networks.

Table 2.6 IEEE and G.hn standards related to PLC

Feature	G.hn	IEEE 1901
Medium	Coax, phone line, power line	Power line
Environment	Home	Home and access
Modulation	FFT-OFDM	FFT- and wavelet-OFDM
FEC	LDPC based	Turbo based and RS based
FEC rates	1/2, 2/3, 5/6, 16/18 and 20/21	1/2, 16/21, 16/18
Subcarrier spacing	195.31 kHz (coaxial), 48.82 kHz (phone lines), 24.41 kHz (power lines)	24.41 kHz
MIMO	Defined in G.hn 9963	Not defined

2.10 Comparison with Wi-Fi

We summarize the main differences between Wi-Fi and PLC technologies on the PHY layer in Table 2.7. Wi-Fi (802.11ax) and PLC (IEEE 1901 FFT-based) use different frequency spectra, with Wi-Fi operating on noninterfering channels for transmission with 20-, 40-, 80- or 160-MHz bandwidth – depending on the 802.11 version – and PLC using the entire available spectrum. Both mediums are multipath, which leads to an OFDM-based PHY layer with PLC allowing each subcarrier to employ a different modulation scheme to tackle high frequency selectivity. Owing to this feature, PLC rate selection is performed at the receiver that has SNR samples over the whole spectrum. In contrast, in state-of-the-art Wi-Fi solutions, the transmitter estimates the optimal data rate, based on the acknowledgments it receives and a moving average of the packet error rate. These differences can lead to potential high overhead for rate selection in PLC, because PLC channel estimation necessitates message passing (sound frames and MMs) per pair of stations. However, compared to Wi-Fi, it can yield more efficient adaptation to channel errors, as we will notice in our experimental study in the next paragraph. The frequency and time selectivity are affected by different factors for the two technologies. This diversity can be beneficial for augmenting network reliability in residential and enterprise settings.

The modulation schemes for PLC are slightly different from those for Wi-Fi, both using QAM-based schemes. The FEC of FFT-based PLC is TCC, whereas Wi-Fi uses LDPC and binary convolutional codes (BCC).

2.11 Summary

In this chapter, we have provided a tutorial on the PLC PHY layer. First, we have identified the main features of the channel to provide a better understanding of end-to-end user performance, which is discussed later in this book. Second, we have pointed out the most popular standardizations of commercial solutions to date. These solutions are designed to overcome challenging channel conditions and provide high data rates and

Table 2.7 Wi-Fi versus PLC: comparison of the main characteristics and features

Feature	Wi-Fi	PLC
Frequency bands	2.4 or 5 GHz	1–86 MHz
Bandwidth	Variable (20/40/80/160 MHz)	Static (depending on HomePlug AV/AV2)
Channel nature	Multipath	Multipath
Interference	Yes	Yes
Frequency selectivity causes	Reflection, diffraction, scattering of signal	Electrical appliances impedance, circuit breakers, phase coupling
Time selectivity causes	Mobility of terminals or scattering objects, fading	Impedance variation and noise introduced by electrical appliances
OFDM	All subcarriers use the same scheme	Each subcarrier may use a different scheme
Modulation	BPSK, QPSK, 16/64/256/1024 QAM	BPSK, QPSK, 8/16/64/256/1024/4096 QAM
FEC	BCC, LDPC	TCC
Rate selection	At transmitter, by using packet error rate	At receiver, with feedback to the transmitter

robust communication. We have observed that PLC design is much more complex than Wi-Fi, and we have explained various intricacies associated with the design of PLC.

References

[1] S. Guzelgöz, H. B. Çelebi, T. Güzel, H. Arslan, and M. K. Mihçak, "Time frequency analysis of noise generated by electrical loads in PLC," in *Proceedings of the 2010 17th IEEE International Conference on Telcommunications (ICT)*, pp. 864–871.

[2] M. Zimmermann and K. Dostert, "A multipath model for the powerline channel," *IEEE Transactions on Communications*, vol. 50, no. 4, pp. 553–559, 2002.

[3] H. A. Latchman, S. Katar, L. W. Yonge, and S. Gavette, *Homeplug AV and IEEE 1901*. John Wiley and Sons, Inc., 2013.

[4] H. B. Çelebi, "Noise and multipath characteristics of power line communication channels," Master's thesis, Department of Electrical Engineering, University of South Florida, Tampa, 2010.

[5] J. A. Cortés, L. Díez, F. J. Cañete, and J. López, "Analysis of the periodic impulsive noise asynchronous with the mains in indoor PLC channels," in *2009 IEEE International Symposium on Power Line Communications and Its Applications (ISPLC)*, pp. 26–30.

[6] S. Sancha, F. Cañete, L. Diez, and J. T. Entrambasaguas, "A channel simulator for indoor power-line communications," in *2007 IEEE International Symposium on Power Line Communications and Its Applications (ISPLC)*, pp. 104–109.

[7] *IEEE Standard for Broadband over Power Line Networks: Medium Access Control and Physical Layer Specifications*, IEEE Standard 1901-2010.

[8] N. Andreadou and F.-N. Pavlidou, "Modeling the noise on the OFDM power-line communications system," *IEEE Transactions on Power Delivery*, vol. 25, no. 1, pp. 150–157, 2010.

[9] Ofcom, "The likelihood and extent of radio frequency interference from in-home PLT devices," Tech. Rep., June 2010, [Online]. Available: www.ofcom.org.uk/_data/assets/pdf_file/0024/35907/pltreport.pdf. [Accessed: November 6, 2019].

[10] B. Zarikoff and D. Malone, "Experiments with radiated interference from in-home power line communication networks," in *2012 IEEE International Conference on Communications (ICC)*, pp. 3414–3418.

[11] *IEEE Standard for Information Technology – Telecommunications and Information Exchange between Systems – Local and Metropolitan Area Networks – Specific Requirements Part 11: Wireless LAN Medium Access Control (MAC) and Physical Layer (PHY) Specifications*, IEEE Standard 802.11-1997.

[12] C. Berrou, A. Glavieux, and P. Thitimajshima, "Near Shannon limit error-correcting coding and decoding: Turbo-codes. 1," in *1993 IEEE International Conference on Communications (ICC)*, pp. 1064–1070.

[13] A. Schwager, "An overview of the HomePlug AV2 technology," *Journal of Electrical and Computational Engineering*, vol. 2013, special issue, 2013.

[14] A. Schwager, W. Bäschlin, H. Hirsch, P. Pagani, N. Weling, J. L. G. Moreno, and H. Milleret, "European MIMO PLT field measurements: Overview of the ETSI STF410 campaign & EMI analysis," in *2012 IEEE International Symposium on Power Line Communications and Its Applications (ISPLC)*, pp. 298–303.

[15] S. Katar, B. Mashbum, K. Afkhamie, H. Latchman, and R. Newrnan, "Channel adaptation based on cyclo-stationary noise characteristics in PLC systems," in *2006 IEEE International Symposium on Power Line Communications and Its Applications (ISPLC)*, pp. 16–21.

[16] *Unified High-Speed Wireline-Based Home Networking Transceivers – System Architecture and Physical Layer Specification*, IITU-T G.9960.

[17] *Unified High-Speed Wireline-Based Home Networking Transceivers – Multiple Input/Multiple Output Specification*, IITU-T G.9963.

[18] *Coexistence Mechanism for Wireline Home Networking Transceivers*, IITU-T G.9972.

3 The MAC Layer of PLC

3.1 Introduction

We present the PLC MAC layer and its most important subfunctions, according to the IEEE 1901 standard. To avoid transmission overheads under large traffic demands, PLC aggregates multiple Ethernet packets into a single PLC frame that is forwarded to the PHY layer for transmission. First, we describe how data is organized in PLC firmware before transmission, and we present the MAC service data unit (MSDU) and MAC protocol data unit (MPDU) creation processes. We then describe the frame aggregation process from a MAC-layer perspective and provide some level of abstraction for facilitating the modeling and simulation of this procedure, without requiring a complex channel model. Frame aggregation concatenates multiple MSDUs into an MPDU, which is the information the PHY layer transports. After the MPDU is created, it is transmitted over the medium with carrier-sense multiple access with collision avoidance (CSMA/CA) or time-division multiple access (TDMA) protocols. CSMA/CA is attempted by the highest-priority stations, based on their traffic type. The priority level is resolved with a protocol in which stations transmit signals in two priority resolution slots. We discuss in detail the CSMA/CA features of PLC, which is the main access control mechanism. This protocol is defined in the IEEE 1901 standard, hence we refer to it as "IEEE 1901." The PLC version of CSMA/CA is much more complex than its Wi-Fi counterpart, but it can offer more potential in terms of performance. TDMA is rarely used in commercial devices. We give a short overview, as described in the standard. MAC-layer performance and TDMA scheduling depend on network management. At the end of this chapter, we explore the network structure, topology, and management methods of PLC.

3.2 MAC Frame Streams and Queue Creation

We start by describing the process of MAC service data unit (MSDU) aggregation and queue creation. The PLC firmware has limited buffers. Yet, it has to support communications with several stations in both management and data modes. PLC creates multiple "MAC frame streams" and queues for links with different destinations or traffic classes. MAC frames that belong to the same stream are forwarded to a unique queue (buffer). The streams are grouped based on the destination, link identifier (priority), and

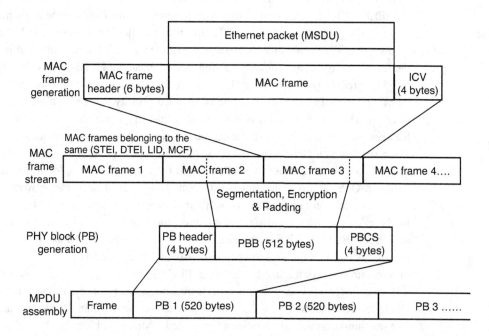

Figure 3.1 PLC MPDU generation. Reproduced with permission from John Wiley and Sons (2013). *Homeplug AV and IEEE 1901: A Handbook for PLC Designers and Users*, © 2013 John Wiley and Sons. All rights reserved.

management/data classification. Buffer management and flow control mechanisms provision resources based on the traffic requirements in terms of throughput, delay, and reliability (quality of service). Owing to bad channel conditions, any part of the frame can be missed at the receiver due to multiple bit errors that cannot be decoded by the FEC module. Frames are fragmented into segments, and each segment is tagged with a sequence number to identify, wait for, and retransmit missed segments. Segments are then aggregated into an MPDU.

Ethernet (MSDU) packets are encapsulated into a MAC frame that contains additional MAC-layer information, as shown in Figure 3.1. The MAC frame header contains the MAC frame type (2 bits) and its length (14 bits). Optionally, this header also includes an arrival timestamp of 4 bytes or a confounder of 4 bytes, which are pseudo-random values for management frames. A confounder is always present for management frames to enhance PLC security. The MAC frame is ended by an integrity check value (ICV), which is a 4-byte cyclic redundancy check (CRC) over the frame's payload. A CRC function enables fast error detection on data segments. CRC calculates a short and fixed-length sequence of bytes (check value or CRC) for each segment of data. When the codeword is received, the device can compare its check value with the new one calculated from the data segment. If the CRC values do not match, then the segment contains error(s).

To describe the MAC frame stream generation, we use some of the terminology from the IEEE 1901 standard and commodity hardware. Any PLC device is identified by a terminal equipment identifier (TEI) within a specific PLC network. TEI is an 8-bit number that identifies a station associated with the local PLC network and is used in the frame control of any unicast transmission. We use the DTEI (destination TEI) and STEI (source TEI) as defined by IEEE 1901 for the destination and source station of an MSDU. Any PLC communication is assigned priority, which is called a "link identifier" (LID). LID is an 8-bit number used to identify a specific allocated connection between two stations. LID represents the priority of a quality-of-service (QoS) frame when it takes values between 0 and 3 (channel access [CA] priorities CA0 to CA3, as described in more detail in Section 3.4), or it can represent a global link identifier with non-QoS data. A global link is established and controlled by the network manager (we give details about the PLC network management in Section 3.7). It can be used for broadcast/multicast and unicast streams, and it has to be provisioned with every station involved (by using management messages for a link handshake process). Every global link has a unique identifier called GLID, which is used to identify different types of allocation. According to the standard, PLC can use contention-based (CSMA/CA) and -free (TDMA) allocations for MPDU transmissions. For contention-free allocation, GLID further identifies the unique link that can use the medium. Broadcast and multicast transmissions are tagged with a flag called a Multicast Flag (MCF). All the above parameters determine the MAC frame stream to which an MSDU belongs.

The PLC MAC aggregates Ethernet packets carrying MSDU payload based on the {STEI, DTEI, LID, MCF} tuple with which they are associated. MAC frames carrying data belonging to the same {STEI, DTEI, LID, MCF} are concatenated to form a MAC frame stream. MAC frame streams that carry data, as opposed to management frames, are referred to as "data streams." The MAC aggregates frames with management information based on the DTEI with which they are associated, thus forming a MAC "management frame" stream. A DTEI of 0xFF is used for a multicast/broadcast transmission. As a result, all connectionless multicast and broadcast frames that belong to the same LID are aggregated into a single MAC frame stream at the transmitter. Figure 3.1 shows an example of chaining multiple MSDUs into a MAC frame stream.

MAC frames are generated from MSDUs, and multiple MAC frames that belong to the same stream are aggregated into MAC frame streams. The MAC frame stream undergoes fragmentation (transmitter) and defragmentation (receiver) processes. It is segmented into 512-byte blocks (can be 136-byte for management), called physical blocks (PBs), which form a MAC protocol data unit (MPDU) payload. A PB corresponds to a PHY-layer FEC block of a physical layer protocol data unit (PPDU). As we discuss in Section 2.4, FEC detects and corrects bit errors, and it does so on a per PB block basis. Hence only the PBs that were not received or were corrupted on the receiver side are retransmitted. MAC framing and fragmentation reduce overheads by maximizing the size of MPDUs transmitted over the medium. The number of MSDUs in an MPDU depends on the link PHY rate (BLE) and the maximum MPDU duration, as we discuss in the next section.

A PB can contain a fraction of at least one MAC frame, depending on the MSDU sizes. For each segment, the MAC layer keeps track of the offset of the first MAC frame boundary within the PB. This information is transmitted with the PB and is used by the DTEI to defragment MAC frames from the received PBs. The MAC frame stream might not have enough bytes to fill the last PB, and in such a case, the last PB can be padded with nondata bits. Once a PB is formed, it represents a single FEC block. The MAC aims to deliver the PB reliably and in order with respect to other PBs. After segmentation, each PB is encrypted using the Advanced Encryption Standard (AES), and information on MPDU header and the location of PB in the MPDU is introduced in its header. We discuss the IEEE 1901 encryption mechanisms in Chapter 7. The PB has a header, encrypted data called the PB body (PBB), and a check sequence (PBCS). The header has the sequence segment number (SSN), which is 16 bits long, and the MAC frame boundary offset associated with the PB. SSNs are used for retransmission of corrupted PBs, and they enable reception of out-of-order PBs and potential duplicate detection at the receiver. The PB header also informs the receiver whether the PB belongs to a data or a management stream. The receiver uses the PBCS field to check the PB's integrity by using a 32-bit CRC.

Ensuring sufficient buffers for multiple links, management, and beacon queues is challenging on PLC hardware. Each MAC frame stream has to keep several states both at transmitter and receiver given the lossy PLC channels. In our experiments, we have found that some commercial devices do not support more than eight simultaneous destinations due to buffer limitations.

3.2.1   MPDU Delimiter

After its generation, the MPDU is transmitted with a preamble and 128-bit frame control (FC) information, followed by the encrypted PBs. The MPDU frame control contains the type of MPDU (aka delimiter type), which can be beacon, Start of Frame (SoF), Selective Acknowledgment (SACK), Request to Send (RTS)/Clear to Send (CTS), or sound frame for tone map estimation (see Section 2.7). For all delimiter types, the first 8 bits of frame control represent (1) the delimiter type (3 bits), (2) the access mode (1 bit), and (3) the short network identifier (SNID) with 4 bits. Table 3.1 presents the fields of frame control for a data MPDU (i.e., SoF). We give guidelines on how to collect and analyze the FC for data MPDUs in Section 4.7.

⚑ Further Reading   We now describe each field of Table 3.1. For the fields of other delimiter types, the reader can refer to the IEEE 1901 standard.

Delimiter Type (DT_IH:) A 3-bit field that classifies the delimiter as beacon (0b000 value), SoF (0b001), SACK (0b010), RTS/CTS (0b011), sound for capacity estimation (0b100), or Reverse SoF (RSoF) (0b101). The rest of the values are reserved for future use. The RSoF is a long sound MPDU that is used to carry both SACK information and payload.

Access Field (ACCESS): A 1-bit field that shows the type of network on which the MPDU is transmitted, that is, in-home (value 0b0) or access (value 0b1) network.

Short Network Identifier (SNID): A 4-bit field that identifies the PLC logical network

Table 3.1 IEEE 1901 start-of-frame fields with the sequence that they are transmitted

Field	Field size (bits)	Definition
DT_IH	4	0b001 (Delimiter Type: Frame, beacon, ACK, RTC/CTS, Sound)
ACCESS	1	Access Field (CSMA/CA or TDMA)
SNID	4	Short Network Identifier
STEI	8	Source Terminal Equipment Identifier
DTEI	8	Destination Terminal Equipment Identifier
LID	8	Link Identifier (priority or global link)
CFS	1	Contention-Free Session
BDF	1	Beacon Detect Flag
HP10DF	1	TIA-1113 Detected Flag
HP11DF	1	1901-aware-TIA-1113 Detect Flag
EKS	4	Encryption key select
PPB	8	Pending PHY Blocks
BLE	8	Bit Loading Estimate
PBSz	1	PHY Block Size
NumSym	2	Number of Symbols
TMI	5	Tone Map Index
FL	12	Frame length
MPDUCnt	2	MPDU Count in burst
BurstCnt	2	Burst Count
BBF	1	Bidirectional Burst Flag
MRTFL	4	Max Reverse Transmission Frame Length
DCPPCF	1	Different CP PHY Clock Flag
MCF	1	Multicast Flag
MNBF	1	Multi-Network Broadcast Flag
RSR	1	Request SACK Retransmission
MST	1	MAC SAP Type
MFSCmdMgmt	3	Management MAC Frame Stream Command
MFSCmdData	3	Data MAC Frame Stream Command
MFSRspMgmt	2	Management medium access control (MAC) frame stream (MFS) Response for the data sent in the preceding Reverse SoF
MFSRspData	2	Data MFS Response for the data sent in the preceding Reverse SoF
BM-SACK	4	Bit Map SACK info for the PBs sent in the preceding Reverse SoF
FCCS	24	Frame Control Check Sequence

between MPDUs transmitted by different networks on the same PLC medium. It is a short representation of the network identifier (NID) and it is similar to service set identifier (SSID) for Wi-Fi APs.

Source Terminal Equipment Identifier (STEI): An 8-bit field that represents the TEI of the station transmitting the MPDU. When repeating a broadcast MPDU, this field is set to the TEI of the original sender, not the repeater.

Destination Terminal Equipment Identifier (DTEI): An 8-bit field that represents the TEI of the destination of the MPDU.

Link Identifier (LID): An 8-bit field for the connection or the priority class to which the MPDU payload is assigned (see MAC frame stream creation in previous section).

Contention-Free Session (CFS): A 1-bit field that shows whether the SoF MPDU is transmitted by contention-free access or CSMA/CA.

Beacon Detect Flag (BDF): A 1-bit field that specifies whether the station had heard the central beacon transmission. In CSMA/CA-only access, this field is reset to 0b0 if it has been more than one beacon period since the last central beacon was received.

TIA-1113 Detect Flag (HP10DF): A 1-bit field that reports any detection of HomePlug 1.0.1 stations. All HomePlug/IEEE 1901 stations that detect a Homeplug 1.0.1 transmission report it to the network manager using this flag to achieve backward compatibility.

1901-aware-TIA-1113 Detect Flag (HP11DF): A 1-bit field that shows whether the station has detected a HomePlug 1.1 (i.e., AV specification) transmission, similarly to the HP10DF flag.

Encryption Key Select (EKS): The 4-bit index of the encryption key used for encrypting the PBs in the MPDU. We give more detail on these values in Chapter 7.

Pending PHY Blocks (PPBs): This 1-byte field informs the receiver about the total number of PBs pending transmission at the completion of the current MPDU, which are previously untransmitted PBs and those awaiting retransmission. PPBs are encoded with a mantissa (M) on the four most significant bits and an exponent (E) on the four least significant bits. The PPB's value can be retrieved using

$$PPBs = (M + 16) \cdot 2^{E-1} + 2^{E-2}, \text{ if } E \geq 1 \quad PPBs = M, \text{ if } E < 1. \tag{3.1}$$

Bit Loading Estimate (BLE): We have discussed BLE in Section 2.7.1. It represents the number of data bits (i.e., data bit prior to FEC) that can be carried on the channel per microsecond. BLE is an 8-bit field encoded using a 5-bit mantissa (five most significant bits) and a 3-bit exponent. BLE value can thus be computed by

$$BLE = (M + 32) \cdot 2^{E-4} + 2^{-5}. \tag{3.2}$$

PHY Block Size (PBSz): A 1-bit field for the size of the PBs in the MPDU, with a value of 0 indicating 520 byte blocks and value of 1 for 136 byte blocks.

Number of Symbols (NumSym): A 2-bit field that represents the number of OFDM symbols used for transmitting the MPDU payload: 0b00 value for no symbols for payload; 0b01 and 0b10 for one and two symbols, respectively; and a value of 0b11 for more than two symbols for the payload.

Tone Map Index (TMI): A 5-bit field that is the tone map identification to be used by the receiver in demodulating the MPDU payload. The receiver saves the tone maps negotiated with the transmitter and uses the TMI to correctly demodulate the MPDU. TMI varies within an AC cycle and is periodic (with six to seven TMIs per cycle).

Frame Length (FL): A 12-bit field that is the duration of MPDU and the interframe space that follows its payload in multiples of 1.28 μs. We give more details on the PLC interframe spaces in Section 3.5.1.

MPDU Count (MPDUCnt): A 2-bit field that is used for MPDU bursts and denotes the number of MPDUs to follow in the current burst transmission. A maximum of four MPDUs are allowed in a burst, hence this field can take values 0–3.

Burst Count (BurstCnt): A 2-bit field that is used to identify an MPDU burst for a particular global link. For local and priority links, BurstCnt is set to 0b00.

Bidirectional Burst Flag (BBF): A 1-bit field that enables a bidirectional burst to continue after the MPDU.

Maximum Reverse Transmission Frame Length (MRTFL): This field is valid only when BBF value is 0b1. It specifies the maximum frame length that a receiver can use in a reverse SoF. Values 0x0–0xF specify multiple lengths from 163.84 µs to 2293.76 µs (see [1]).

Different CP PHY Clock Flag (DCPPCF): A 1-bit field that is set to 0b1 when a station uses the PHY receive clock correction of a different network manager during the contention period, and not the network manager of the SNID indicated in the SoF. Stations have to synchronize their PHY clocks based on beacons.

Multicast Flag (MCF): A 1-bit field that shows multicast/broadcast (value 0b1) or unicast data (value 0b0) for the MPDU.

Multi-Network Broadcast Flag (MNBF): A 1-bit field to show that the MPDU is a broadcast to all stations regardless of the SNID. Only a single PB containing management data is allowed in the MPDU when using this mode.

Request SACK Retransmission (RSR): A 1-bit field that recovers a missing SACK MPDU on global links.

MAC SAP Type (MST): A 1-bit field that indicates the MAC service access point (SAP) for which the current MPDU payload is designated. SAP is the type of higher protocol used for the MSDU and has a value 0b0 for Ethernet and a 0b1 value reserved for future use.

Management MAC Frame Stream Command (MFSCmdMgmt): A 3-bit field that represents a command from the transmitter's management MAC frame stream to the corresponding reassembly stream at the receiver. Examples of commands for the receiver's buffers are "INIT," "SYNC," "RE_SYNC," "RELEASE."

Data MAC Frame Stream Command (MFSCmdData): A 3-bit field that represents a command from the transmitter's data MAC frame stream to the corresponding reassembly stream at the receiver. Examples of commands for the receiver's buffers are the same as for the MFSCmdMgmt.

Management MAC Frame Stream Response (MFSRspMgmt): A 2-bit field with the response from the receiver's MFSCmdMgmt command in a reverse SoF. Examples of responses are "ACK," "NACK," "FAIL," "HOLD".

Bit Map SACK Information (BM-SACKI): The field acknowledges the PBs transmitted in the previous reverse SoF. The least significant bit holds the reception status of the first PB, and so on.

Frame Control Check Sequence (FCCS): A 24-bit field that contains the CRC-24 value computed over the fields of the frame control. The standard specifies the CRC function. FCCS is used to check the integrity of the SoF.

3.3   Physical Blocks Aggregation and Retransmission

Frame structure and aggregation are important for the end-to-end performance of PLC, especially for overhead reduction and reliable transmission. To reduce various types of overhead, multiple Ethernet packets are transmitted within a single PLC frame. The

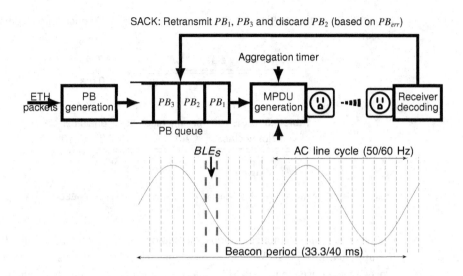

Figure 3.2 The PB aggregation and acknowledgment process in PLC.

number of Ethernet frames (MSDUs) varies per MPDU, with a complex process that depends on the BLE adaptation and prior retransmissions of PBs. It is possible to understand and simulate the whole process with a few MAC metrics, avoiding the complexity of exact channel and modulation modling.

From a performance and simulation perspective, the metrics of interest of the frame aggregation process are the number of Ethernet packets per MPDU, the MPDU duration, and the PB error rate PB_{err}. Figure 3.2 sketches the MAC-layer processes that determine these metrics. Ethernet packets are fragmented into PBs. Then, the PBs are forwarded into a queue, and based on the BLE of the current tone map slot s BLE_s, they are aggregated into an MPDU. The MPDU duration is computed based on BLE_s, on the maximum MPDU duration (specified by the IEEE standard [1]), and on an aggregation timer that can fire every few hundreds of microseconds after the arrival of the first PB. The MPDU duration must be a multiple of the OFDM symbol length, and padding can be used to fill these symbols. The aggregation timer limits the maximum time to wait for the Ethernet packet for aggregation and can reduce delay for small packet transmissions. MPDU is then converted to a PPDU via the PHY layer. The modulated PPDU is transmitted by a CSMA/CA protocol described in the next section. The receiver demodulates and decodes each individual PB of the MPDU and transmits a SACK that informs the transmitter about which PBs were received with errors, so that only the corrupted PBs are retransmitted. If a frame is received with all the PBs corrupted, then it is assumed to have collided with another frame. An MPDU has a time-to-live (TTL) and is discarded if its transmission is not over within TTL ms. TTL depends on the priority class, as we discuss in Section 3.4. In case an MPDU reaches the maximum frame duration and there is high traffic intensity, multiple MPDUs can be transmitted. In this case, stations are allowed to transmit multiple MPDUs in a burst via CSMA/CA access. Bursts contend for the medium and not individual MPDUs. The number of MPDUs per

Table 3.2 Default HomePlug AV TTL per channel access (CA) priority from commercial
chipsets

Priority class	CA0/CA1	CA/CA3
TTL	2 s	300 ms

Table 3.3 IEEE 1901 priority resolution signals per class

Priority class:	CA0	CA1	CA2	CA3
PRS0			✓	✓
PRS1		✓		✓

burst depends on channel conditions and station capabilities. Up to four MPDUs may
be supported in a burst. We can observe that the full retransmission and aggregation
process, and consequently the MAC and PHY layers, can be emulated and modeled using the two metrics PB_{err} and BLE_s and the maximum MPDU and burst limits defined
by the standard. Experimental measurements of these metrics can be plugged into any
simulator along with an assumed PB error distribution for PB_{err}. Thus an exact channel representation per link and a PHY layer implementation can be circumvented for
MAC-layer PLC simulations.

Retransmission is handled with an ARQ protocol. ARQ is implemented on a per
MAC frame stream basis, requesting retransmissions when there are errors on transmitted PBs due to bad channel conditions or collisions. The receiver stores in a buffer the
correctly received PBs for in-order delivery, in a memory that is shared between the
different data flows. The transmitter resends the corrupted PBs one or more times in the
following MPDUs, according to the received SACK. A detailed buffer management and
state machine for retransmissions can be found in the IEEE 1901 standard [1].

3.4   Link Priority and Priority-Resolution Process

The PLC MAC layer uses four priority classes for best-effort and delay-sensitive applications, which are CA0 to CA3, from lowest to highest priority. The CA0/CA1 classes
have different CSMA/CA parameters compared to CA2/CA3. For example, the standard
specifies a TTL on the frame transmission that is vendor specific. MPDU is discarded
if transmission is not successful within TTL. The default value of most commercial
chipsets of our experiments was found to be 2 s for the CA0/CA1 class and 0.3 s for the
CA2/CA3, as shown in Table 3.2. TTL is configurable, and we give the guidelines on
how to modify TTL in Chapter 4.

PLC frame priority depends on the application, the source and destination of the
frame, and the higher networking layer's priority, such as the type-of-service field of
the IP layer or the VLAN priority field in the IEEE 802.1p virtual LAN standard. In
Section 4.7, we explain how to modify the PLC priority of a flow or how to assign

application-based priorities. The standard enforces that only the stations with traffic that belongs to the highest priority can contend for the medium through the CSMA/CA process described in the next section. The highest priority is found using two consecutive priority-resolution time slots, called PRS0 and PRS1, in which a station with packets of a specific priority class either transmits a specific priority signal or senses the power line medium for a higher-priority signal. Stations with data of highest priority CA3 transmit on both PRS0 and PRS1. CA2 stations transmit a priority signal only on PRS0, and CA1 stations transmit only on PRS1 and only when they sense that PRS0 is empty. With this process, the stations sense signal transmissions in PRS0 and infer whether a station with higher priority aims to transmit, in which case, they defer their transmission in PRS1 or in CSMA/CA contention. Table 3.3 summarizes the priority-resolution signal transmissions. The priority-resolution slots are allocated automatically every successful transmission in the channel.

The PLC priority-resolution scheme can result in a complete starvation of best-effort services (CA0/CA1). Some of our experiments have shown that when traffic demands for CA2/CA3 classes are high, CA0/CA1 classes get starved and do not receive channel access. Such performance degradation has also been observed by Cano and Malone [2]. It is important to properly configure the network and the TTL values for the different classes to harmonically coexist.

3.5   CSMA/CA

The MAC layer of IEEE 1901 specifies both TDMA and CSMA/CA protocols [1]. However, most commercial in-home devices implement only CSMA/CA. The same CSMA/CA protocol is used for all PLC specifications described in Table 2.2. The first HomePlug specification that uses this CSMA/CA mechanism is HomePlug 1.0. HomePlug 1.0 uses a frame preamble that comprises 7.5 OFDM symbols. In CSMA/CA, the station has to sense the medium to decide if it is busy or idle for a fixed duration called the time slot. The slot duration was decided by the time required by a station to determine whether the medium is busy or idle, that is, to detect a preamble transmission, in HomePlug 1.0. Preamble in 1.0 had a duration of 38.4 μs. Although newer technologies have different symbol and preamble durations, the slot duration has remained the same for all HomePlug standards for backward compatibility. We compare the IEEE 1901 CSMA/CA with the one of 802.11, which is similar but simpler.

The backoff process of IEEE 1901 uses the backoff counter (BC) and the deferral counter (DC). In addition, there are four backoff stages. In the standard, the backoff stage is determined by the so-called backoff procedure counter (BPC) [1]. When a new PPDU arrives for transmission, the station starts at backoff stage 0, and it draws the backoff counter BC uniformly at random in $\{0, \ldots, CW_0 - 1\}$, where CW_0 is the contention window for backoff stage 0. Similarly to IEEE 802.11, a 1901 station decreases BC by 1 at each time slot if the station senses the medium to be idle, and it is frozen when the medium is sensed busy. If the medium is sensed busy, BC is also decreased by 1, once the medium is sensed idle again. When BC reaches 0, the station attempts to transmit the PPDU. Also similarly to IEEE 802.11, a 1901 station jumps to the next backoff stage and increases BPC, if the transmission fails (unless it is already at the last backoff stage, in which case it reenters this backoff stage). Upon entering backoff stage i, a station draws BC uniformly at random in $\{0, \ldots, CW_i - 1\}$, where CW_i is the contention window at backoff stage i, and the process is repeated. After a successful transmission, backoff stage is reset to 0. For IEEE 802.11, the contention window is doubled between two successive backoff stages, thus $CW_i = 2^i CW_0$. For IEEE 1901, CW_i depends on the priority level (CA0 to CA3, as described in Section 3.4) and is given in Table 3.4. The backoff processes of 1901 and 802.11 are presented as a finite state machine in Figure 3.3.

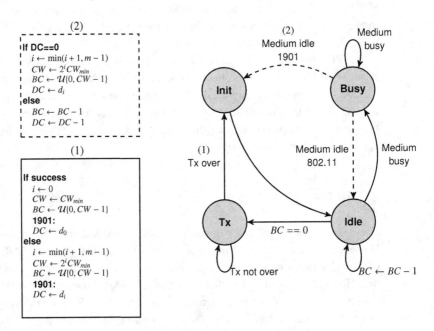

Figure 3.3 A finite-state machine that represents the 802.11 and 1901 backoff procedures, as running in a station at each time slot. A station is in state **Init** whenever it selects BC (or updates DC in 1901); in states **Busy** and **Idle** when the medium is sensed busy and idle, respectively; and in state **Tx** when it is transmitting. After a transmission attempt, stations move to state **Init** and execute Algorithm (1). In 1901, after choosing a new BC, the station updates DC according to the backoff stage i. The parameter m denotes the total number of backoff stages. The dashed lines highlight the fundamental difference between the two standards. After sensing the medium idle while in **Busy** state, 1901 stations move to state **Init** and execute Algorithm (2). In contrast, 802.11 stations return to the **Idle** state and just resume BC.

When there are few contending stations (i.e., one or two), or when the traffic load is very low, the time spent in backoff causes a large overhead and increases as the contention window increases. Given the large slot duration of the IEEE 1901 standard, the average delay due to backoff ($[CW_0 - 1]/2$ time slots) can be reduced when there are few contending stations – hence a small collision likelihood – by choosing a small CW_0, for example, $CW_0 = 8$, as specified for IEEE 1901 (Table 3.4). However, smaller CW

Table 3.4 IEEE 1901 parameters for the contention windows CW_i and the initial values d_i of deferral counter DC, for each backoff stage i

Priority class	CA0/CA1		CA2/CA3	
Backoff stage i	CW_i	d_i	CW_i	d_i
0	8	0	8	0
1	16	1	16	1
2	32	3	16	3
3	64	15	32	15

causes higher collision probabilities when the number of stations increases or when the traffic load rises. The probability that two or more stations select the same backoff counter increases as CW decreases. The deferral counter DC can help as a counter-measure in the CSMA/CA process of IEEE 1901, to reduce collisions caused by small contention windows. This is achieved by triggering a redraw of the backoff counter BC even before the station attempts a transmission.

When entering backoff stage i, DC is set at an initial DC value d_i, where d_i is given in Table 3.4. After having sensed the medium busy, a station decreases DC by 1 (in addition to BC). If the medium is sensed busy and $DC = 0$, then the station jumps to the next backoff stage (or reenters the last backoff stage, if it is already at this stage), and it redraws BC without attempting a transmission. In other words, if a station senses the medium busy more than d_i times before its transmission at backoff stage i, then the station moves to backoff stage $i + 1$ without transmitting. Figure 3.4 shows an example of an IEEE 1901 backoff process with two stations.

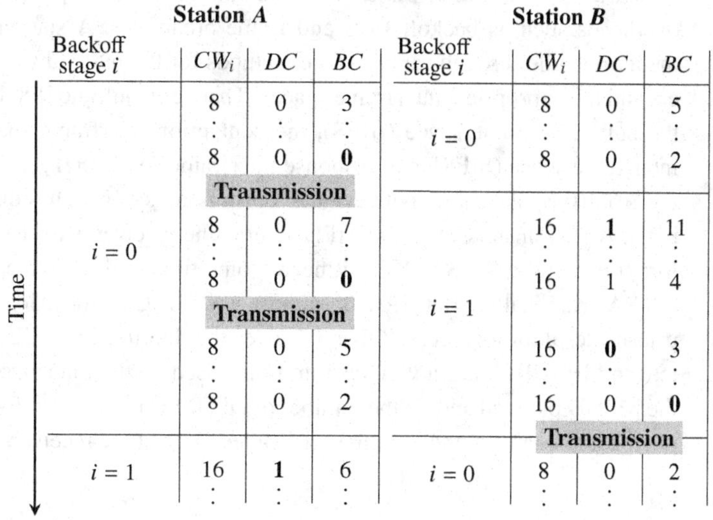

Figure 3.4 Evolution of the IEEE 1901 backoff process with two saturated stations A and B. Initially, both stations are at backoff stage 0. A transmits twice consecutively. Backoff stage i changes when a station senses the medium busy and has $DC = 0$.

We have pointed out the use of the deferral counter to reduce collisions based on sensed channel activity. We now discuss the effect of the deferral counter on delay, since this can explain the rest of the CSMA/CA parameter choices in IEEE 1901. Owing to the deferral counter, the transmitting station has an advantage over the other stations because its CW is reset at its initial smallest value CW_0, whereas the other stations might increase their CW when $DC = 0$, as explained earlier. As a result, a station can win the channel for multiple consecutive transmissions, until it releases the channel; then it can wait for a long sequence of other station transmissions, until it obtains the medium again, which can result in high delay variance called jitter. To reduce the CW imbalance between the transmitting station and the others, IEEE 1901 uses four backoff stages, allowing at most three times for the CW to be doubled. Moreover, best-effort and delay-sensitive applications use different configurations of parameters (see Table 3.4). The delay-sensitive class (CA2/CA3) uses smaller contention windows, and the contention window is not doubled between backoff stages 1 and 2. This can reduce jitter but can yield a higher collision probability, that is, lower throughput, compared to the CA0/CA1 classes. Figure 3.5 summarizes the design implications of the MAC parameters described in this section.

3.5.1 Interframe Spaces

An interframe space (IFS) is a time gap between frames or acknowledgments on the PLC medium. Interframe spaces are needed between data transmission and acknowledgment for the receiver to have sufficient time to decode the data and to encode the acknowledgment. One of the reasons for frame aggregation used in PLC and Wi-Fi standards (802.11n/ac/ax) is to avoid additional IFS on a per packet basis, and other overheads such as backoff slots and frame preambles. A station determines that the medium is idle through carrier-sense methods for the interval specified by frame headers and the appropriate interframe spaces. There are multiple IFS defined in IEEE 1901: the burst interframe space (BIFS), the contention interframe space (CIFS), the short interframe space (SIFS), and response interframe space (RIFS). As we observe in Figure 3.6, BIFS takes place between the transmissions of MPDUs that belong in the same burst. After transmission of an MPDU burst, the receiver waits for an RIFS interval before transmitting the SACK, as it needs some time to decode the last PBs and prepare the SACK. Finally, after SACK, stations wait for CIFS duration before restarting the priority resolution process. Table 3.5 gives the durations of IEEE 1901 MAC parameters and IFS. RIFS has a variable duration, based on the tone map used for modulating the MPDU payload and on the number of OFDM symbols used for carrying the MPDU payload, to enable sufficient time for PB decoding at the receiver.

3.5.2 Physical and Virtual Carrier Sense

Clear channel assessment (CCA) is the procedure that determines whether the medium is busy. PLC deploys two methods for CCA (similarly to Wi-Fi): physical carrier sense

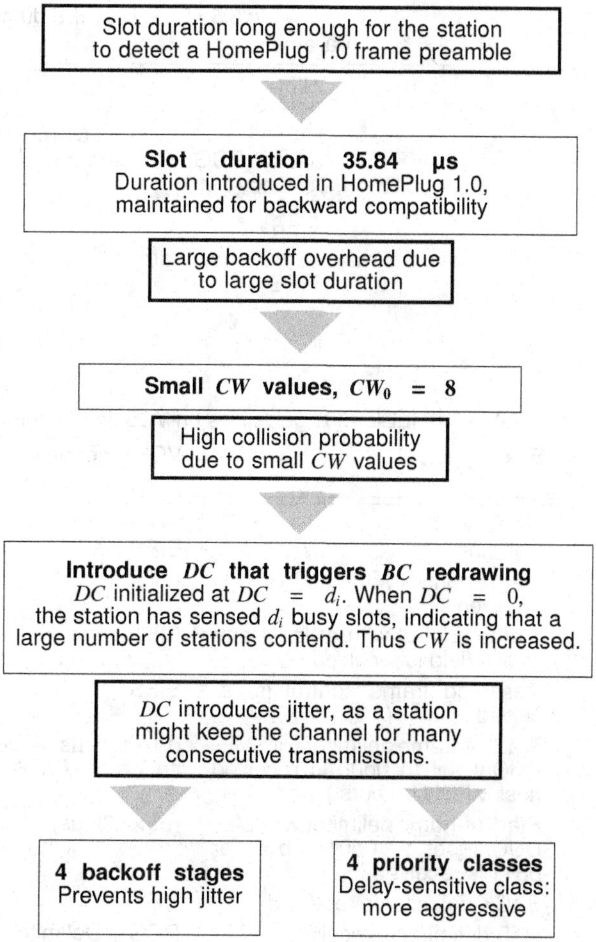

Slot duration long enough for the station to detect a HomePlug 1.0 frame preamble
Slot duration 35.84 μs
Duration introduced in HomePlug 1.0, maintained for backward compatibility
Large backoff overhead due to large slot duration
Small CW values, CW0 = 8
High collision probability due to small CW values
Introduce DC that triggers BC redrawing
DC initialized at DC = di. When DC = 0, the station has sensed di busy slots, indicating that a large number of stations contend. Thus CW is increased.
DC introduces jitter, as a station might keep the channel for many consecutive transmissions.
4 backoff stages
Prevents high jitter
4 priority classes
Delay-sensitive class: more aggressive

Figure 3.5 A discussion of the design implications of the IEEE 1901 CSMA/CA parameters.

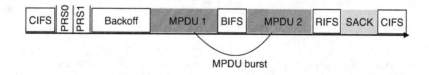

CIFS
PRS0
PRS1
Backoff
MPDU 1
BIFS
MPDU 2
RIFS
SACK
CIFS
MPDU burst

Figure 3.6 PLC interframe spaces, priority resolution slots, and MPDU bursting.

(PCS) and virtual carrier sense (VCS). PCS works on the PHY layer by detecting signals such as the priority resolution or frame preamble. If PCS detects a signal, then the backoff counter described in the previous section is paused and PCS stays active until a full preamble is detected and decoded by the PHY/MAC layers. After decoding a frame preamble, the VCS comes into play on the MAC layer: the PLC station tags the medium

Table 3.5 IEEE 1901 IFS durations

Parameter	Duration
Slot	35.84 µs
PRS	35.84 µs
RIFS	30–160 µs (140 µs by default)
RTS/CTS Gap (RCG)	120 µs
CTS-MPDU Gap (CMG)	120 µs
CIFS	100 µs
BIFS	20 µs
MaxFL (maximum frame length)	2501.12 µs
EIFS	2920.64 µs

Table 3.6 IEEE 1901 guidelines on VCS timer after certain events in the channel

Event type	VCS timer value	Medium state when VCS timer expires
Collision	EIFS	Idle
Frame control with bad FCCS is received	EIFS	Idle
Frame control with at least one invalid field is received	EIFS	Idle
Reserved frame control is received	EIFS	Idle
Start-of-frame delimiter with MPDUCnt set to 0b00 is received (last MPDU on burst)	$(FL \times 1.28\ µs)$ + DelimiterTime + CIFS	PRS0
Start-of-frame delimiter with MPDUCnt set to 0b01, 0b10, or 0b11 is received	$(FL \times 1.28\ µs)$	Search for next MPDU in the burst
SACK delimiter is received	CIFS	PRS0
RTS delimiter is received	RCG + DelimiterTime + CMG + DelimiterTime	Idle
CTS delimiter is received	$(DUR \times 1.28\ µs)$ + CIFS	PRS0

busy for an expected duration of channel occupancy, determined by the duration field in the decoded frame preamble. The station maintains the network allocation vector (NAV) for tracking the busy channel periods. If the preamble cannot be decoded (e.g., due to collisions), the station sets NAV to a maximum value called extended interframe space (EIFS). The station does not attempt any frame transmission until its NAV expires. Table 3.6 summarizes the values of VCS depending on the sensing events over the PLC channel. The values of VCS depend on whether the device is able to decode the frame control. The frame control contains the frame length (FL) and the position of the MPDU in the burst (MPDUCnt). If the device is not able to decode at least one of the fields of frame control, the frame is considered as collided.

3.6   Additional Access Methods

3.6.1   TDMA

According to the standard, a beacon period can consist of both contention (CP) and contention-free (CFP) periods. CP access is enabled with the CSMA/CA protocol, whereas CFP access relies on TDMA, which is managed by the CCo. To gain access to CFP, an IEEE 1901 station must ask for a time allocation from the CCo by transmitting a management message in a CP. This management message is called CC_LINK_NEW.request and contains the MAC addresses of the source and destination of data, a traffic specification, and the BLE of both directions of the communication link. CCo provides schedule information (i.e., CP and CFP time allocations for stations) in beacons. The standard proposes data streaming, such as video, audio, and Voice over Internet Protocol (VoIP) to be transmitted during a CFP. Our lab experiments show that CFP and TDMA are not supported in commodity PLC hardware. We explain how to detect MPDUs transmitted via CFP/TDMA in Section 4.7.3.

3.6.2   RTS/CTS

RTS/CTS are control packets that can be transmitted before actual data. The purpose of such control packets is to protect the transmission from hidden-terminal effects, where two contending stations A and B cannot hear each other, but an intermediate client C can listen to both A and B without being able to decode packets under simultaneous transmission of A and B (see Figure 3.7). By transmitting an RTS, the destination station C can reply with a CTS if it senses the channel to be idle. CTS is overheard by both A and B, hence setting their NAV vector and eliminating the possibility of collision. In case of a collision of RTS packets, the CTS is not sent, and no data is transmitted. Typically, RTS collisions are much shorter than data frame collisions, as RTS packets contain very few bytes. Hence RTS/CTS saves airtime in the medium and alleviates long collisions due to hidden terminals.

   RTS/CTS is provisioned by the IEEE 1901 standard to solve hidden terminal issues. However, RTS/CTS is an optional mechanism for in-home systems, as it can introduce overheads when hidden terminals do not exist, especially for short packet transmissions. It can harm delay-sensitive applications. During our lab experiments, we observe that commodity devices do not typically use RTS/CTS. We give details on how to detect RTS/CTS packets in Section 4.7.3.

3.7   MAC Layer and Network Management

Similarly to Wi-Fi access points, a PLC station is responsible for transmitting beacons and authenticating/associating stations that join the network. This station is called the central coordinator.

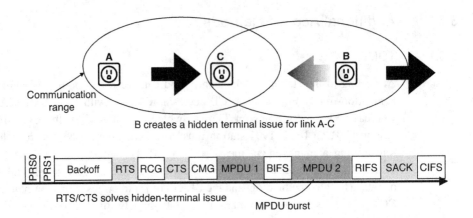

Figure 3.7 PLC control frames and interframe spaces for hidden terminals.

3.7.1   Central Coordinator

PLC has a centralized authority, called the central coordinator (CCo), that manages the network. Each PLC station must join a network with a CCo; however, transmissions between stations do not need to go through the CCo. Typically, one of the first stations plugged becomes the CCo. The CCo can change dynamically if another station has better channel conditions than the current CCo. The CCo associates and authenticates new stations and allocates TDMA and CSMA/CA sessions over the beacon period, if both protocols are used. Allocation of TDMA and CSMA/CA regions is necessary, so that TDMA transmissions, which do not perform sensing, do not collide with CSMA/CA. It also transmits beacons periodically, so that the stations' PHY clocks are synchronized with the AC line cycle. The beacons are important for both PHY and MAC layers, as explained in Section 3.7.3. The CCo is responsible for the coordination with neighboring networks and for the topology discovery. Each CCo creates a logical network, which is an abstract network that defines possible data connection links between stations and is independent of the physical location of each device in the network. Only stations on the same logical PLC network are allowed to communicate with each other. Different logical networks can coexist on the same medium with collaboration of their CCos for TDMA allocations and by carrier sensing under CSMA/CA transmissions. Within a logical network, transmitted frames are encrypted using the network membership key (NEK) that is unique per logical network and can change dynamically. A station needs a network membership key (NMK) to be part of a logical network and get the NEK. Hence there might be logical networks in the same physical medium, but they may not be able to communicate because they have different NMKs. We describe the authentication and encryption processes in Chapter 7.

The CCo is updated automatically in case of failure, through standard-defined management messages. Thus the user does not have to configure a CCo or handle CCo changes. HomePlug implementations provide seamless network operation and hide the CCo complexities from the user. All stations behave the same from end user, higher layer, and application perspectives, whether they act as a CCo or not. In Section 4.5, we give guidelines on how to set the CCo statically in the case where a constant network structure is desired, such as in enterprise or large residential environments.

The CCo learns the topology of its PLC network and of any other neighboring networks. Each station can broadcast a "discover beacon" periodically (at slots allocated by the CCo). The discover beacon has information about the station and its logical network. Each station that overhears the discover beacon adds its information to a discovered station list (DSL) if the station belongs to the same network or to a discovered networks list (DNL) if the station transmitting the discover beacon belongs to another logical network. The CCo can request from each station its DSL and DNL to create a topology map. The CCo can use this topology to determine if there is another station in the network that would be a better CCo than itself. The criteria for making this decision, in order of priority, are

- user selection;
- CCo capability;
- number of discovered stations in the DSL; and
- number of discovered networks in the DSL.

Every station in the logical network has to communicate directly or indirectly with the CCo. In case some stations are not able to hear the CCo transmissions, the standard provisions an optional mechanism to set up a proxy CCo for these stations. The proxy CCo creates a proxy network that includes the stations hidden to the original CCo. The proxy CCo periodically transmits beacons by forwarding the beacons of the original CCo with CSMA/CA access and at highest priority class (CA3).

3.7.2   Network Topology

A CCo has similar functionalities to the access point (AP) in the infrastructure mode of Wi-Fi (the most popular Wi-Fi mode). However, the network topologies with PLC and with Wi-Fi in infrastructure mode differ: in-home PLC typically works in a mesh structure, that is, any two nodes communicate with each other directly without packets being transmitted through CCo. In contrast, with the Wi-Fi infrastructure mode, two nodes cannot communicate directly with each other and have to transmit through the AP. Therefore, the network topology in Wi-Fi infrastructure mode is a star, whereas the topology with PLC is a mesh. Regarding data transmissions, PLC is similar to the mesh mode of Wi-Fi, where an ad-hoc network is formed. Another difference between CCo with PLC and AP with Wi-Fi is that CCo can change dynamically to another device due to harsh channel conditions, whereas the Wi-Fi AP is usually preconfigured and rarely changes.

Similarly to Wi-Fi, a PLC network has a unique identification called the short network identifier (SNID). The equivalent Wi-Fi term would be the SSID. SNID is a 4-bit field in the IEEE 1901 frame control that is used to distinguish between MPDUs transmitted by different networks on the same PLC channel. The SNID field is present in all MPDU delimiters, including SoF, SACK, sound ACK, RTS/CTS, data MPDU, and sound MPDU. SNID is a short version of the 54-bit network ID (NID) that uniquely identifies a logical network. The NID changes only when the network membership key changes. We describe how a new NID is generated in Chapter 7.

3.7.3   Beacons

Similarly to the AP in the infrastructure mode of Wi-Fi (the most popular Wi-Fi mode), the CCo in a PLC network transmits beacons periodically for network discovery and synchronization (see Section 3.7.3). The beacons are used to synchronize the stations to the mains cycle, to inform about TDMA and CSMA/CA allocations within the beacon period, and to manage the network. Management and vendor-specific information can be also transmitted within the beacons. The beacon interval is twice the period of the mains cycle, hence it is 40 ms for 50-Hz AC frequency and 33.3 ms for 60 Hz. The beacon is usually transmitted within a beacon region that follows every second positive zero-crossing of the AC line cycle, as shown in Figure 3.8. The beacon is transmitted at a fixed offset Δ from the zero-crossing of the AC line cycle, within a beacon region reserved for beacons. The beacon region can have several beacons in case multiple PLC networks are within sensing range of each other, and it enables multiple beacons to be transmitted without collision. The CCo is responsible for synchronizing the beacon transmission to the AC line cycle, which is achieved by setting the CCo to track a particular point in the AC line cycle using a digital phase locked loop (DPLL) and to eliminate noise events and jitter in the measurement of the line cycle phase. The CCo uses its local tracking history to predict future locations of the beacons and announces these locations in the beacon.

The beacon delimiter fields are the 4-bit SNID; a 32-bit beacon timestamp, which represents the CCo's network time base (NTB) at the start of the corresponding beacon PPDU; and beacon transmission offsets BTO (0 to 3), which are used to report the offset of subsequent beacons from their expected locations based on the CCo's 25-MHz clock. BTO fields, together with a resource allocation schedule, enable a station to transmit accurately in its allocation when a beacon is missed. The beacon MPDU has a single PB payload of 136 bytes, modulated using a robust modulation scheme (see Table 2.3). A beacon can be one of the three types defined by IEEE 1901: central, transmitted by the CCo; discover, transmitted by any associated and authenticated station periodically to assist the network-topology discovery; and proxy, transmitted by a station set as a proxy CCo to manage hidden stations within the same network. The beacon payload has multiple fields, such as the 54-bit NID; the IEEE 1901 FFT hybrid mode, which is a 2-bit field that shows whether the network is in coexistence mode with prior PLC standards; the STEI of the CCo transmitting the beacon; the beacon type (e.g., central, proxy, or discover); any uncoordinated networks reported (UCNR), a 1-bit field that informs about the absence or presence of networks that are not coordinating with the CCo; and the network power-saving mode (NPSM), a 1-bit field that informs the stations whether the CCo has configured the network into a power-save mode. The detailed beacon payload fields can be found in the standard [1].

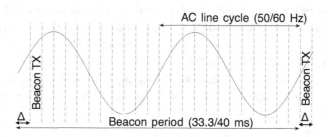

Figure 3.8 Mains cycle and beacon interval.

3.7.4 Security and Authentication

Power line signals can propagate from one cable to another due to electromagnetic coupling. Therefore, privacy and security issues are of concern with PLC. To address such issues, PLC uses security mechanisms built on 128-bit advanced encryption standard (AES). Stations have to be configured with a network membership key (NMK), which acts similarly to the password for security in Wi-Fi. Only stations that share a common NMK can decrypt each other's data.

The CCo ensures security and privacy in the network by associating and authenticating stations. During the association procedure, a station obtains a TEI from the CCo. It then provides the station with a network encryption key (NEK), as long as the station has a valid NMK, which is the authentication phase. The station encrypts each PB with the NEK. We describe how to set the NMK of a station in Section 4.5. We present the association and authentication processes in Chapter 7.

3.7.5 Management Messages

Management messages (MMs) are a key feature for PLC. They are used for network management, authentication, association, and tone map adaptation. Stations must exchange MMs each time the tone map is updated, because the source has to be notified of the modulation scheme that each carrier uses and other PHY parameters. If MMs are transmitted through the PLC channel, they are tagged with high (CA2 or CA3) priority, since they are important for the network operation. In addition to being transmitted between stations over the power line medium, MMs can be transmitted from the Ethernet interface host to the PLC chipset. This enables users to interact with and configure the devices or measure statistics, as we explain in the next chapter.

The format of MMs is based on the standard Ethernet frame format, with a unique EtherType assigned to IEEE 1901. Different MMs are identified by a field called MMType. MMType is a 2-octet field that defines the MM that follows in the payload. The two least significant bits of MMType indicate whether the message is a request, confirmation, indication, or response. The three most significant bits of the MMType represent the category to which the MM belongs (e.g., between station and CCo, between proxy CCo and CCo, between stations, or vendor specific). MMs can be standardized or be specific to vendor implementations. The standardized MMs are described in detail in IEEE 1901 standard [1]. The vendor-specific MMs are used to enhance the functionality of proprietary systems when exchanged between stations of the same vendor. The first 3 bytes of the vendor-specific MM contain the IEEE-assigned OUI (organization unique identifier) as described in the IEEE 802-2001 standard. The remaining payload data in these MMs are defined by the vendor. We describe such MMs in more detail in Section 4.2.

3.8   MAC-Layer Improvements of HomePlug AV2

The MAC layer of HomePlug AV2 introduces improvements to prior HomePlug AV. The main enhancements are new power-save modes, short delimiters, delayed acknowledgments, and immediate repeating.

The power-save mode of AV2 is first introduced in HomePlug Green PHY, as Green PHY devices are often resource constrained or battery powered. Power save is important given the large number of smart and automation devices in residential and enterprise environments. In power-save mode, stations limit their power consumption by periodically moving between awake and sleep states. The stations can only transmit and receive packets in the awake state and during a period called the awake window (from a few milliseconds to multiple beacon periods). Similarly, the sleep window is the period during which the station does not transmit or receive frames. HomePlug power-save modes are implemented with a power-saving schedule (PSS), which defines the awake windows and the duration between consecutive awake windows, called the power-save period (PSP). PSP is restricted to multiples of the beacon period. Figure 3.9 shows an example of the PSS for four stations A–D. HomePlug AV2 enables each station to have a different PSS. As a result, all stations need to have the PSS of the other stations in the network, since PLC enables station-to-station communication. In the example of the figure, Stations C and D can only communicate every four beacons with other stations. The CCo is responsible for determining the PSS per station, enabling or disabling power-save mode for the whole network, and waking a station in power-save mode. IEEE 1901 defines MMs that the CCo can send to stations for configuring PSS [1]. In commercial PLC devices, power-save mode is triggered after a few minutes (e.g., 5 minutes [3]) of inactivity and can also be disabled by the user.

New short delimiters and delayed acknowledgments of AV2 aim to reduce MAC-layer overhead and increase the effective throughput. The duration of the HomePlug AV delimiter is 110.5 μs and can cause significant overhead for each PPDU channel access. The delimiter first contains a preamble, which is used to detect the beginning of the packet and to estimate the channel, so that the data can be decoded. Then, the delimiter includes the frame control, which is the MPDU SoF. HomePlug AV2 has a new delimiter that can limit the overhead associated with delimiters by reducing the duration to 55.5 μs and to a single OFDM symbol. In particular, it combines preamble and frame control in the same OFDM symbol: every fourth subcarrier in the OFDM symbol is used as a preamble carrier, and the other carriers are used for the frame control. Figure 3.10

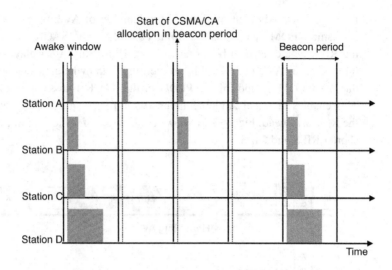

Figure 3.9 HomePlug AV2/GreenPHY power save example. Reproduced from *An Overview of the HomePlug AV2 Technology*, Hindawi Publishing Corporation. © 2013 Larry Yonge et al.

shows how AV2 transmits preamble and frame control simultaneously. In addition to delimiter reduction, the RIFS and CIFS interframe spaces described in Table 3.5 are reduced to 5 and 10 µs, respectively. This enables further overhead reduction from IFS. The disadvantage of a new short delimiter is that it cannot be detected asynchronously in CSMA/CA transmissions, hence the receiver needs to know when the short delimiter begins. As a result, the short delimiter is only used for SACK, reverse SoF, and TDMA MPDUs, where the receiver is aware of the time arrival of the delimiter. Backward compatibility is ensured by appropriately configuring the frame length FL in the MPDU SoF.

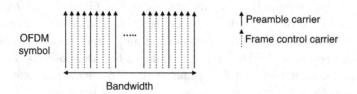

Figure 3.10 HomePlug AV2 short delimiter example over the OFDM subcarriers.

The short RIFS can create a timing issue of decoding the last PB and preparing the SACK at the PLC receiver. The time required to decode the last OFDM symbol of the frame and to prepare the SACK can be high on PLC hardware, thus requiring a sufficiently large RIFS. In HomePlug AV, since the preamble is a fixed signal separate from the frame control, the preamble portions of the acknowledgment can be transmitted while the receiver is still decoding the last OFDM symbol and encoding the payload for the acknowledgment. With the short delimiter of AV2, the preamble is encoded in the same OFDM symbol as the frame control for the SACK (see Figure 3.10), thus the RIFS would need to be larger than for HomePlug AV. Delayed acknowledgment introduced in AV2 resolves this problem by acknowledging the segments ending in the last OFDM symbol of the PPDU in the SACK transmission of the next PPDU. This enables practical implementations with a very small RIFS, reducing drastically the RIFS overhead. Figure 3.11 shows the benefits of a short-delimiter ACK and much shorter RIFS and CIFS.

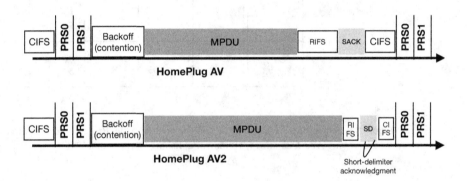

Figure 3.11 HomePlug AV2 MAC-layer enhancements: overheads are drastically reduced by short-delimiter acknowledgments and much shorter CIFS and RIFS.

Repeating can extend network coverage by using multihop routes and is also defined as an optional mechanism in IEEE 1901 in-home networks. A repeater is a commodity PLC station that repeats the PLC signal to the final destination. HomePlug AV2 expands coverage and addresses hidden station issues by allowing to repeat the MPDU on paths with better channel quality than a single hop link. Figure 3.12 shows repeating between two stations A and B. A two-hop communication should be targeted for links when single-hop capacity is worse, for example, when $\min(BLE_{A-R}, BLE_{R-B}) > BLE_{A-B}$ holds for the BLE of each link, since the throughput of two hops is bounded by the slowest link. Repeating in AV2 is done using a single channel access, hence there is a single end-to-end acknowledgment from the final destination of the MPDU. MPDU segments are not stored at the repeater to minimize latency and buffer occupancy. Thus reassembly and segmentation are not required at the repeater.

3.9   HomePlug GreenPHY

HomePlug GreenPHY uses certain features from the IEEE 1901 standard and introduces new functions for low power consumption, scalability, and high reliability, while ensuring interoperability with other HomePlug and 1901 technologies. Table 3.7 summarizes the GreenPHY functions that are different to the technologies described in previous sections. GreenPHY simplifies PLC design by using only the three ROBO modes presented

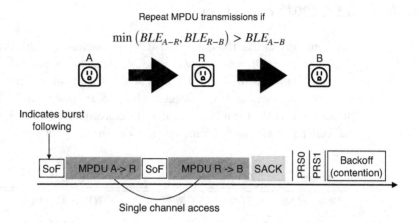

Figure 3.12 HomePlug AV2 single-access repeating.

in Table 2.3. As a result, it also uses only 1/2 FEC rate and QPSK modulation. Puncturing in the Turbo interleaver is not supported in GreenPHY (see Figure 2.6). By using the ROBO preknown modulation schemes, tone map estimation and adaptation (see Section 2.7) are not needed in a GreenPHY network, which simplifies device design and reduces multiple overheads. Although GreenPHY devices do not initiate tone map estimation, they are required to respond to any other station initiating sounding. ROBO modes can be demodulated by any HomePlug or 1901 device.

On the MAC layer, GreenPHY only uses CSMA/CA, as opposed to 1901 devices, that can be designed to support both TDMA and CSMA/CA. Although TDMA connections cannot be controlled by GreenPHY devices, a GreenPHY device can participate in a network of AV devices that use TDMA allocations. GreenPHY reuses the priority resolution process described in Section 3.4. It also includes new power-save modes (see Figure 3.9) and new entries in the CCo beacon to support power-save modes and to communicate those to other devices. GreenPHY stations do not transmit bursts, but they can receive them, in order to defer transmissions to other HomePlug devices. Finally, to enable multihop repeating, new functions and MMs are provisioned for the CCo.

Table 3.7 HomePlug GreenPHY main features on PHY/MAC layer

Feature	HomePlug GreenPHY
Frequencies	2–30 MHz
Tone maps	Only ROBO
FEC rate	Only 1/2
Modulation	Only QPSK
Access method	Only CSMA/CA
Power-save	New functions and MMs
Beacon	New entry for power save
MPDU Bursting	No bursts transmission; can receive bursts
Repeating	Adds proxy CCo and routing functions

3.10   IEEE 1901 versus G.hn

The G.hn 9961 standard has similar MAC-layer functions to IEEE 1901 [4, 5]. Table 3.8 summarizes the differences between the two standard families. As we observe, both standards enable CSMA/CA and TDMA access. Quality of service is supported in both, by contention-based and -free access. RTS/CTS is optional in both G.hn and IEEE. Bursts in IEEE 1901 can be either uni- or bidirectional. However, G.hn only enables bidirectional frame bursts. Frame aggregation is used by both standards.

Table 3.8 IEEE and G.hn standards related to PLC

Feature	G.hn	IEEE 1901
Channel access	TDMA, CSMA/CA	TDMA, CSMA/CA
Quality of Service	Supported	Supported
RTS/CTS	Optional	Optional
Burst modes	Bidirectional	Uni-/bidirectional
Frame aggregation	Supported	Supported
Management	Medium-access plan (MAP)	Beacon

The difference between the two standards lies in the resource allocation and management functions. A G.hn network is managed by a domain manager (DM). Contrary to IEEE 1901, which is managed by beacons, a G.hn network is managed by a medium access plan (MAP) transmitted by its DM. The MAP contains the contention-free (TDMA) allocations and a group of one or more stations to contend in the CSMA/CA time allocation. This is different from IEEE 1901, where any station can contend for the CSMA/CA period of the beacon cycle. Hence, with appropriate optimization on the G.hn DM, the collision probability in G.hn can be lower than that in IEEE 1901 networks. A CSMA/CA period in G.hn is divided into time slots. Because a DM can assign the same time slot to a number of stations, a station performs a backoff procedure upon receiving a shared time slot prior to channel access to avoid collisions. When a station obtains a time slot designated for access, it transmits an "INUSE" signal and starts the priority-resolution procedure. G.hn uses a similar PRS concept to IEEE 1901: a station resolves the priority level before the backoff state. Other stations that are not assigned to access the slot stop time slot counting as soon as they receive the INUSE signal. A MAP message by the DM is used for the next MAP cycle, whereas a beacon from the CCo is used for the current beacon cycle.

3.11   Comparison with Wi-Fi

We summarize the main differences between Wi-Fi and PLC technologies on the MAC layer in Table 3.9. Wi-Fi uses channelization and allows for multiple noninterfering channels for transmission with 20-, 40-, 80-, or 160-MHz bandwidth – depending on the

802.11 standard. In contrast, PLC uses the entire available bandwidth, which depends on the device specification (e.g., HomePlug AV vs. AV2). The most popular topology is mesh for PLC and star for Wi-Fi in infrastructure mode. Wi-Fi typically uses a static AP for management, whereas the CCo for indoor PLC networks can change dynamically. Both PLC and Wi-Fi are subject to interference due to the broadcast nature of the medium, and to reduce collisions, they use CSMA/CA protocols.

Table 3.9 Wi-Fi (802.11a/g/n/ac standards) versus PLC on the MAC layer of residential and enterprise networks

Feature	Wi-Fi	PLC
Channelization	Yes (20/40/80/160MHz)	No
Network structure	Infrastructure (transmissions must pass though access point) or ad-hoc mode	Central coordinator (CCo) obligatory, but transmissions do not need to go through CCo
Management	AP (static)	CCo (can change dynamically)
Medium nature	Shared (broadcast)	Shared (broadcast)
Slot duration	9 or 20 µs	35.84 µs
CSMA/CA	Doubles CW only after collision	Doubles CW after collision and possibly also after a sensed transmission

The IEEE 1901 PLC CSMA/CA protocol is similar to the 802.11 counterpart. The main difference is that, contrary to 802.11, 1901 stations increase their CW not only after a collision but also after sensing the medium busy, regulated by the deferral counter. This results in IEEE 1901 having two distinct features for channel access control:

- With IEEE 1901, the contention window can be increased without suffering from a collision, hence a station can adapt its transmission aggressiveness to the level of contention in the network, without wasting channel time in collisions.
- By appropriately configuring its deferral counter, IEEE 1901 can dynamically adjust its level of reaction to contention with a fine granularity; in contrast, with IEEE 802.11, the level of reaction cannot be tuned.

Figure 3.13 provides potential gains of 1901 versus 802.11. We show the time evolution of the protocol dynamics and compare 1901 with 802.11 for two different values for the number of stations in the network, $N = 2$ and $N = 15$. For $N = 2$, 802.11 has a large average CW only after consecutive collisions, whereas 1901 adapts its CW even after successful transmission attempts. The 1901 average CW increases or remains constant as a station keeps transmitting and decreases only when the other station successfully wins the medium. For $N = 15$, 1901 adapts its CW before a collision, thus it wastes less time in collisions, and the average CW converges much faster to the steady state average CW. Hence, under dynamically changing traffic, 1901 can adjust its CW to the load demand faster than 802.11.

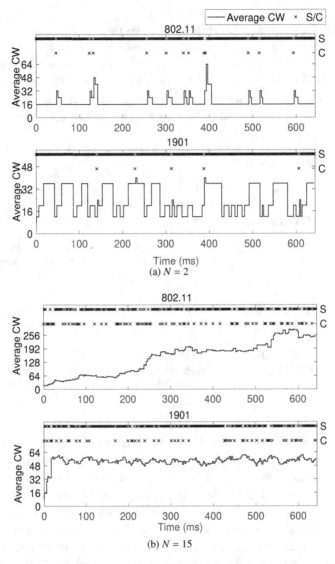

Figure 3.13 Time evolution of the contention window CW averaged over all stations in the network (left axis) and the binary outcome of transmission attempts (right axis), i.e., success (S) or collision (C), in simulation. All stations are saturated and start at backoff stage 0; 1901 is simulated with CA1 class and 802.11 with $CW_0 = 16$ and 7 backoff stages (i.e., 802.11a/g/n).

3.12   Summary

In this chapter, we described the main features of the PLC MAC layer. We have explained all subfunctions that transport Ethernet packets to the PHY layer for transmission. First, packets are grouped into MAC frame streams depending on their destination and priority. MAC frame streams enable efficient buffer management and aggregation of multiple packets into a single MPDU for transmission. The MPDU obtains access to the channel depending on its priority and on the multiple-access protocol. We have presented both the popular CSMA/CA protocol and the less deployed TDMA access specified by IEEE 1901. Finally, we have underlined the main MAC-layer management functions of PLC and the features of HomePlug AV2.

References

[1] *IEEE Standard for Broadband over Power Line Networks: Medium Access Control and Physical Layer Specifications*, IEEE Standard 1901-2010.

[2] C. Cano and D. Malone, "When priority resolution goes way too far: An experimental evaluation in PLC networks," in *2015 IEEE International Conference on Communications (ICC)*, pp. 952–957.

[3] tp-link, "How to enable/disable power saving mode on powerline adapters?" Tech. Rep., [Online]. Available: www.tp-link.com/us/faq-2018.html. [Accessed: November 6, 2019].

[4] *Unified High-Speed Wireline-Based Home Networking Transceivers – Data Link Layer Specification*, IITU-T G.9961.

[5] M. M. Rahman, C. S. Hong, S. Lee, J. Lee, M. A. Razzaque, and J. H. Kim, "Medium access control for power line communications: An overview of the IEEE 1901 and ITU-T G. hn standards," *IEEE Communications Magazine*, vol. 49, no. 6, pp. 183–191, 2011.

4 Experimental Framework

4.1 Introduction

Commercial PLC devices allow the user to modify certain configurations and parameters and to measure various statistics. These features are often necessary for guaranteeing reliable performance, full coverage, and security in PLC networks. In addition, they can be employed for troubleshooting network issues or improving performance. As explained in the previous chapter, power line is a noisy medium, and the achievable throughput on power line links varies with space and time. Hence there is a need for measuring the link quality to determine the locations for establishing good and stable link performance. Moreover, power lines create a broadcast medium where PLC devices interfere with each other and where their data transmissions might collide. For this reason, MAC-layer measurements and efficient configurations are crucial for PLC. Finally, PLC networks are managed by the central coordinator (CCo), and communication is encrypted for security and privacy. Users and PLC designers should configure these features while setting up their network. In this chapter, we provide guidelines to measure and configure the aforementioned features. Moreover, we show how to develop custom software tools for PLC devices for research, hybrid network design, and optimization.

As discussed in the previous chapter, the IEEE 1901 standard provisions management messages (MMs) to configure the network and to enable coordination between the stations. These MMs are also employed by the user to interact with the PLC devices. In this chapter, we explain how to use MMs and certain open source tools to measure statistics and configure commercial devices. The majority of these MMs are vendor specific. Yet, we have found that they apply to tens of vendors and that they can be used with conventional PLC devices. We focus on the most popular chipsets sold by Intellon/Atheros/Qualcomm. Most of the open source tools have been developed for these chipsets. Then, we show that the same principles can be applied to other vendors, taking an example from other popular chipsets developed by Broadcom. The rest of the chapter is organized as follows.

In Section 4.2, we present in detail the tools we use in this book for managing commercial devices. We discuss the MM structure, two open source tools that operate with Linux devices, commercial tools for other operating systems, and an example of coding new tools for PLC. Next, we discuss in Section 4.3 the firmware properties of state-of-the-art devices and the available firmwares in our testbed. The first guidelines in

Section 4.4 consider modifying the parameter information block (PIB) file of the PLC interface, a file containing information on the device and configurable parameters. We then give more guidelines in Section 4.5: we describe how to set the membership key, set the CCo, and retrieve the PLC network structure. In Sections 4.6 and 4.7, we provide guidelines for retrieving statistics on the PHY- and MAC- layer performance. Finally, in Section 4.8, we show how to develop tools for PLC devices from scratch. This will be very useful later, in Chapter 8, where we design heterogeneous IEEE-1905.1-compliant networks, consisting of Wi-Fi and PLC technologies. All the aforementioned tools are tested with different chipsets from Intellon/Atheros/Qualcomm. In Section 4.9, we explore another major chipset vendor and the up-to-date PLC devices offering 2-Gbps PHY rates. We show that we can apply the same principles of MMs to diverse devices and vendors. We present guidelines and packet traces that enable the user to infer the MM structure for other vendors.

4.2   Hardware and Tools

PLC devices allow limited hardware interaction with the user. The hardware settings and the firmware code are currently inaccessible, but certain firmware parameters can be modified or retrieved. In this section, we describe the hardware used for the tests and guidelines of this book and the state-of-the-art in PLC hardware. Then, we explain in detail how to interact with the firmware, and how to measure statistics with PLC.

4.2.1   Devices and Testbed

The tools presented in this chapter have been tested on a testbed of Alix 2D2 boards from PC Engines running the OpenWrt Linux distribution [1, 2].[1] The boards are equipped with a HomePlug AV miniPCI card (Meconet interface, Intellon INT6300 chip) which interacts with the kernel through a Realtek Ethernet driver, and with an Atheros AR9220 wireless interface.

In addition to using the main testbed, we experiment with commercial HomePlug AV and AV500 devices, such as the Devolo AV200 (INT6300 chip) and the Netgear XAVB5101 (Atheros QCA7400 chip). Overall, all results and guidelines of this chapter have been tested with more than six vendors (such as Netgear, Devolo, TP-Link). The devices deploy either the Intellon INT6300 chipset or the Atheros QCA7400 chipset [3, 4]. In the rest of this chapter, we provide guidelines for both these two hardware series and we refer to them as INT6xxx and QCA7xxx, because the tools have different functions for the two types of hardware.

This study is mainly focused on Intellon/Atheros/Qualcomm chipsets. However, in Section 4.9, we extend the guidelines to other chipset vendors, such as Broadcom. For

---

[1] OpenWrt is a lightweight, GNU/Linux-based operating system for embedded devices, such as residential gateways and routers. Its components have been optimized for size, to be small enough to fit into the limited storage and memory available in routers.

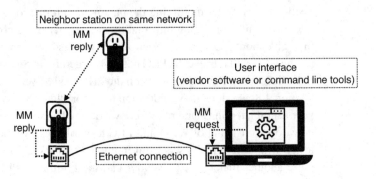

Figure 4.1 The interaction with PLC devices via MMs.

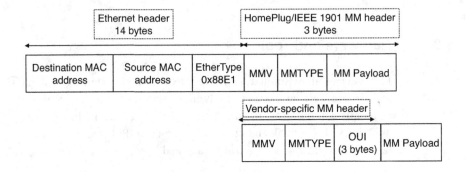

Figure 4.2 The structure of PLC management messages.

these tests, we use HomePlug AV2 devices D-Link DHP-701AV, which support 2-Gbps PHY rates with an effective throughput of 500 Mbps.

4.2.2   Interaction with Management Messages

PLC uses management messages for coordination between the stations or the CCo. In addition, MMs are used for interacting with the PLC device. There are both standardized (with IEEE 1901) and vendor-specific MMs. The standardized MMs are mostly the MMs transmitted between the stations and the CCo over the power line, whose details can be found in the IEEE 1901 standard [5]. Here we focus on the MMs that can be sent to the PLC devices by the end Ethernet host. The user and network administrator can use such MMs to modify the configuration of the PLC device and to retrieve performance statistics; these MMs are mostly vendor specific.

The user's operating system perceives the PLC device as an Ethernet interface. To interact with the PLC chipset, typical vendor software or command-line tools send MM from the Ethernet interface to the PLC chipset, as we show in Figure 4.1. Any configuration or measurement request is transmitted as a specific MM. The PLC chipset replies

Table 4.1 The header fields of an MM transmitted to and from the PLC chip

Field	Value
Destination MAC Address	6-byte MAC address of the destination
Source MAC Address	6-byte MAC address of the source
VLAN Tag (optional)	Variable-length, optional VLAN tag for IEEE 802.1Q standard
EtherType	2 bytes that show which protocol is encapsulated: 0x88E1
MMV	1-byte field for management message version
MMType	2-byte field that determines the type of the MM
OUI	3-byte organizationally unique identifier

Note. The source and the destination of the MM can be either the Ethernet interface or the PLC chip, which have different MAC addresses. Vendor-specific MMs always include the Organizationally Unique Identifier (OUI) for the vendor, enabling unique identification.

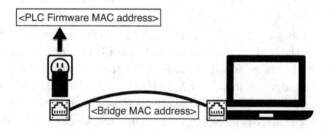

Figure 4.3 MAC address bridging with PLC. The PLC firmware and the Ethernet interface have different addresses. The PLC firmware is used as a bridge to the Ethernet interface, referred to as higher layer entity (HLE) in the standard.

with an MM that contains either the success of configuration or the measurement requested. A lot of software and command-line tools enable also to configure or measure statistics from remote devices that belong in the same network. In such cases, the MM can be forwarded in the PLC medium, and a reply from the remote device is generated.

To configure and manage PLC devices, it is necessary to create and send the correct MMs. The MM should be set to the IEEE-assigned EtherType value of 0x88E1 (HomePlug AV). This 2-byte field indicates which protocol is encapsulated in the payload of the Ethernet frame. Figure 4.2 and Table 4.1 present the structure of the MM, including its Ethernet header. The source MAC address is the MAC address of the user's Ethernet interface. The destination MAC address can be either broadcast (FF:FF:FF:FF:FF:FF) or MAC address of the PLC chip. It is important to mention that the PLC chip has an internal MAC address that is different from the MAC address of the Ethernet interface (which could be retrieved by the ifconfig Linux command). The PLC device is bridged to an Ethernet MAC address, as we show in Figure 4.3. In Section 4.5.3, we will explain how to retrieve the PLC chip MAC address.

The Ethernet header of the MM is followed by the HomePlug AV header, with fields

Table 4.2 IEEE-1901 standardized MMs that can be used with state-of-the-art-devices

MMType	Functionality
0x0014	Central Coordination Discover List Request
0x6020	Get Bridge Infos Request
0x6038	Get Network Infos Request
0x6048	Get Network Stats Request

Table 4.3 The two least significant bits of the MMType indicate the nature of the management message

MMType Two LSB Value	Description
0x00	Management Message Request
0x01	Management Message Confirm
0x10	Management Message Indication
0x11	Management Message Response

such as the MM version and type, and by the payload. The Management Message Version (MMV) is a 1-byte field that indicates the specification version used to interpret the MM. In in-home OFDM-based messages, this field should be set to 0x01. Management Message Type (MMType) is a 2-byte field that defines the structure of the MM payload.

The MMs can be standardized by IEEE 1901 or be vendor specific. All standardized MMs can be found in IEEE 1901 specification [5]. Most of them are sent between PLC devices; in Table 4.2, we show the standardized MMs that can be sent by the user to the PLC device. The two least significant bits of the MM indicate the purpose of the management message, that is, request, confirm, indication, or response. The exact mapping for these four classes can be found in Table 4.3.

Vendor-specific MMs always begin the payload with the OUI of the vendor, which uniquely identifies the vendor. The OUI is the first 3 bytes of the station's MAC address. Table 4.4 presents some MMs that can be used for Intellon and Qualcomm chipsets and for the hardware described in the previous section. All these MMs are used by the open source and commercial tools that we discuss next.

As we observe in Table 4.4, the least significant bit of the MMType of all MM requests is always 0. After the reception of a certain MM request from the station (e.g., 0xA000 for getting the device version), the PLC chip replies with an MM response with the same MMType as that of the request, but with the least significant bit set to 1 (i.e., 0xA001 in our example). For instance, the response to MMType 0x0014 would have an MMType 0x0015. All the aforementioned MMs are important for configuring and measuring statistics over the network.

4.2.3   faifa

faifa has been developed by Florian Fainelli, Nicolas Thrill, and Xavier Carcelle [6]. The tool can configure any Intellon-based Power Line Communication device using Intellon INT5000 and INT6000 series chips (e.g., 6000 and 6300 chips). It supports Intellon-specific and standardized MMs. Such MMs were presented in the previous section.

Table 4.4 Examples of vendor-specific MMs that can be used with Intellon and Qualcomm chipsets

MMType	Functionality
0xA000	Get Device/SW Version Request
0xA004	Write MAC Memory Request
0xA008	Read MAC Memory Request
0xA00C	Start MAC Request
0xA01C	Reset Device Request
0xA020	Write Module Data Request
0xA024	Read Module Data Request
0xA02C	Get Watchdog Report Request
0xA030	Get Link Statistics Request
0xA034	Sniffer Mode Request
0xA038	Network Info Request
0xA040	Check Points Request
0xA048	Loopback Request
0xA04C	Loopback Status Request
0xA050	Set Encryption Key Request
0xA054	Get Manufacturing String Request
0xA058	Read Configuration Block Request
0xA068	Get Device Attributes Request
0xA06C	Get Ethernet PHY Settings Request
0xA070	Get Tone Map Caracteristics Request

faifa works with Linux-based operating systems. For example, it is available for Ubuntu, Red Hat and Debian systems, and there are frameworks for cross-compiling this tool for OpenWrt. To run faifa, the user has to type the commands:

```
1  faifa -m -i <ETH_INTERFACE>
```

The user has to specify the interface of the PLC device <ETH_INTERFACE> (e.g., "eth1" or "eth2"). After running this command, the user can select between a number of requests to send to the PLC interface, such as resetting the device, getting information on the firmware version, updating the security keys, and retrieving statistics for the PHY and MAC layers. In Sections 4.6 and 4.7, we show how to obtain PHY/MAC performance metrics with faifa. The tool is very useful, as it can capture PLC MPDU frame control, and hence provide statistics on a per-MPDU basis. In Section 4.7.3, we show how to enable and use this functionality.

4.2.4   Open PLC Utils

Open PLC Utils has been developed by Atheros/Qualcomm [7]. It includes a comprehensive list of Linux commands with which the user can configure both the Intellon INT6xxx chipset and Atheros QCA7xxx chips. The tool is regularly updated, supports the newest chipsets, and includes extensive documentation. It supports various operating systems, such as Linux, OpenWrt, OpenBSD, MAC OSX, and Microsoft Windows.

By navigating over the source code [7], the user can find the functionality and structure of vendor-specific MMs. Compared to faifa, Open PLC Utils is a more complete tool and operates with state-of-the-art devices, whereas faifa might fail to interoperate with new-generation PLC chipsets QCA7xxx. Open PLC Utils has various

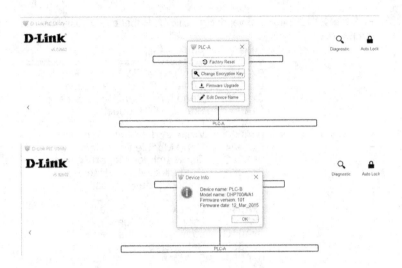

Figure 4.4 A screenshot from the vendor's software capabilities and the PLC device info it can provide.

commands for different functionalities and operations, whereas with `faifa`, the user runs the initial command as described above (`faifa -m -i <ETH_INTERFACE>`), and then all functionalities are offered by the menu that shows up. Each command of `Open PLC Utils` provides a different functionality, and there are several commands for different chipsets. To configure INT6xxx series chipsets, the commands `int6kxxx` must be used, and to configure QCA7xxx chipsets, the `ampxxx` commands must be used. Below, we show examples of PLC device reconfigurations with `Open PLC Utils`.

4.2.5   Commercial Tools

PLC vendors usually offer their own tools to configure their devices. Every PLC device usually comes with a software tool for Windows and other operating systems. These tools can be easily installed, and they include a graphical user interface (GUI).

Figure 4.4 provides a screenshot from a D-Link software utility for PLC. Using this utility, the user can retrieve information about the capacity of the links, the network structure, and the firmware version of the device. Moreover, the user can perform certain actions, such as resetting the device, and configuring security and quality-of-service. Because a list of such software tools from different vendors can be long and not all vendors provide configuration capabilities to the user, we next provide guidelines on how to configure PLC devices with open-source tools that only depend on the firmware chip and not on the vendor of the PLC device. In Sections 4.8 and 4.9, we also show how to extend these tools for chipsets not currently supported by open source tools.

4.2.6   Custom Tools

All open source PLC tools use vendor-specific or IEEE-1901-standardized MMs to interact with the hardware. In this chapter, we show how to develop your own tools for PLC devices and how to interact with PLC hardware using a programming language (e.g., C/C++, Python) that fits your experimental framework. Our example uses C++ and a modular router that has been extensively used in academia and industry, called `Click!` [8]. With the `Click!` modular router, we can create classes called elements to send and receive these MMs. Typical examples that we implement using `Click!` are (1) capacity estimation (using the MMs similar to the `int6krate` command) and (2) capturing packets by enabling the sniffer mode of PLC interfaces. To develop custom tools, we need to know the structure of the MM request and reply. We can capture the MMs by using `tcpdump` or Wireshark [9] to understand their structure, to infer their MMType (field that defines the message type), and to code the `Click!` elements (C++ classes that process packets). Alternatively, the descriptions of these MMs are available through the source code of the tools we are using [7] or IEEE 1901 standard. We show how to write your own tools for PLC devices in Section 4.8.

4.3   Firmware

Compared to certain Wi-Fi drivers, the firmware of PLC is generally obfuscated or encrypted in most cases. This means that modifications to the code are not usually possible and that certain parameters cannot be configured by the user. The firmware consists of two different parts: a .PIB file (Parameter Information Block) and a .NVM file (the firmware image and the code). In the PIB, there are certain parameters, such as the metadata (MAC address, device name, etc.) and the default tone mask, which are discussed in the next section. We can interact with the firmware through MMs. In particular, the version of the firmware can be retrieved and the firmware can be updated with a newer version provided by the vendor.

It is possible to know which version of the firmware is currently used with MMs. The firmware version can be retrieved with `faifa` and `Open PLC Utils`. With `faifa`, the firmware version can be retrieved as follows:

```
1  Get Device/SW Version Request
2  Choose the frame type (Ctrl-C to exit): 0xA000
3  Frame: Get Device/SW Version Request (0xA000)
4
5  Dump:
6  Frame: Get Device/SW Version Confirm (A001), HomePlug-AV
       Version: 1.0
7  Status: Success
8  Device ID: Unknown, Version: MAC-QCA
       7420-1.0.0.345-04-20120507-FINAL, upgradeable: 0
```

With `Open PLC Utils`, the firmware version can be retrieved as follows:

```
1  # Example with INT6xxx hardware
2  int6k -i <ETH_INTERFACE> -a
```

```
3  eth2 00:B0:52:00:00:01 Fetch Device Attributes
4  eth2 00:0B:3B:79:2D:E7 INT6400-MAC
     -3-3-3348-00-2764-20080808-FINAL-B (16mb) (0xF5) "50Hz" "
     Detected"  0xA9
5
6  # Example with QCA7xxx hardware
7  int6k -i <ETH_INTERFACE> -a
8  eth2 00:B0:52:00:00:01 Fetch Device Attributes
9  eth2 28:C6:8E:B6:E6:5E QCA7420-MAC-1-4-20120507-00-0000-
              --? (1mb) (0xF5) "50Hz" "Detected"  0x00
```

If we look at the example with the int6k command and INT6xxx chipset above, we can retrieve the major firmware version 3-3 (see see line 4, text *MAC-3-3*), the PIB version code 3348, the build number 00, the update version 2764, the build date 20080808, and the SDRAM configuration code. Vendors usually provide firmware updates and the software utilities to upload the firmware on the PLC interface. To upgrade a remote device, both PIB and NVM should be updated at the same time. The version codes given as examples above must match those written in the PIB file. Notice that the device reports the AC frequency detected. This is 50 Hz, as our experiments are conducted in Europe.

4.3.1   Modification and Update

The user can update the firmware with Open PLC Utils. With the command amptool (or int6k, depending on the PLC chip, as explained in Section 4.2.4) and, respectively, options -n, -N, the firmware file <NVM_FILE> can be, respectively, downloaded and uploaded:

```
1  amptool -i <ETH_INTERFACE> -n <NVM_FILE> <MAC_ADDRESS> #
     read firmware
2  amptool -i <ETH_INTERFACE> -N <NVM_FILE> <MAC_ADDRESS> #
     write firmware on device
```

The file has a .nvm extension. <MAC_ADDRESS> should be replaced by the MAC address of the PLC interface or simply by the word "local," which indicates the PLC chip connected to the Ethernet interface. To use the commands for a remote device connected on the same PLC network, the MAC address should be replaced appropriately.

4.4   Parameter Information Block (PIB) File

The PIB file enables the modification of the firmware and of variable values that reside in the external nonvolatile random access memory (NVRAM), typically a flash memory. In a typical configuration, PLC interfaces boot from this NVRAM at start-up, receiving operating instructions and parameters. The PIB structure varies across chipsets and examples of this structure can be found at the code of Open PLC Utils.[2] Unfortunately,

---

[2] https://github.com/qca/open-plc-utils/blob/master/docbook/pib.h.html.

the PIB structure can change across chipsets and devices, so one has to infer the PIB structure and the offset of the desired parameters for modification. We give guidelines on how to modify the PIB even when its structure is not known.

4.4.1 Modification

The PIB file can be easily retrieved and modified. The simplest way to download the file from the PLC interface is by using Open PLC Utils. We present the commands to download and upload a modified PIB file to the NVRAM. The PIB can be read in hex format by using the command setpib -v <PIB_FILE> <OFFSET>, at any required offset. Through knowledge of the PIB structure (or trial and error of various offsets), the user can modify the MAC address of the PLC interface, the security keys, the central coordinator of the network (CCo), certain QoS parameters (packet priorities and time-to-live), a static network configuration, and the default OFDM tone mask. As the PIB structure varies per chipset generation, we give in Section 4.5 the guidelines to modify the main network-wide features, such as CCo and security; later in this section, we also show an example of modification of PHY/MAC layer parameters. After any modification of the PIB file, it is recommended to reboot the device.

To download and update a PIB file named <PIB_FILE> with AR7xxx chipsets, the user can run the following commands:

```
1  plctool -i <ETH_INTERFACE> local -p <PIB_FILE>  # Read the
     PIB file from the PLC interface
2  ------ # Modify  <PIB_FILE>
3  plctool local -i <ETH_INTERFACE> -P <PIB_FILE> # Write the
     PIB file to the PLC interface
```

For INT6xxx chipsets, plctool should be replaced by int6k. The user can simultaneously update the firmware and the PIB:

```
1  ampboot -i <ETH_INTERFACE> -N <NVM_FILE> -P <PIB_FILE>
```

We can restore the PIB file to factory defaults. This permanently erases all PIB changes made since the device was last programmed with factory default settings. The device will automatically reset and reboot:

```
1  plctool -i <ETH_INTERFACE> -T
```

4.4.2 Example with MAC-Level TTL

We now give an example of how the PIB file can be useful for modifying certain PHY-/MAC-layer parameters. With our example, we modify the time-to-live (TTL) of PLC frames in their buffer. This TTL parameter varies across different priority classes. The reader can refer to Sections 3.3 and 3.4 for the detailed explanation of TTL and the priority classes. TTL values have to be carefully set, as they affect the end-to-end PLC performance. We next present the Open PLC Utils commands that modify the TTL with QCA7xxx and INT6xxx chipsets.

We start with QCA7xxx chipsets. With the first command (line 2 below), we retrieve the PIB file from the PLC interface. With the second command (line 4), we read the PIB file around offset 700 (in hexadecimal writing) bytes, where we know that the 4-byte TTL values are saved.[3] The PIB file is shown by lines of 16 bytes (32 hexadecimal symbols), with the byte offset from the beginning of the PIB file indicated at the beginning of each line. The default TTL values are 300 000 μs for CA2/CA3 classes or 04 93 E0 in hexadecimal. Owing to the network-byte order conversion, we observe this hexadecimal number reversed as E0 93 04 00 (line 14, starting at offset 710). Similarly, the 2 000 000 μs of priorities CA0/CA1 results in 80 84 1E 00 on the PIB file (line 12, starting at offset 704). With the third command (line 19), we modify the PIB file data at offset 704, which is the TTL for CA0 class. In our example, we set the TTL to 8 000 000 μs (which is 7A 12 00 in hexadecimal, and this corresponds to 00 12 7A 00 converted into network-byte order). Finally, with the fourth command (line 32), we upload the modified PIB file on the PLC interface. With these four commands, we replace the timeout of CA0 with 8 000 000 μs:

```
1   plctool -i <ETH_INTERFACE> local -p <PIB_FILE> #read pib
2
3   setpib -v <PIB_FILE>  700 skip 0 #read the bytes in hex
      format to determine offset
4
5   Result:
6
7   000006E0 00 00 00 00 00 00 00 00 00 00 00 00 00 00 00 00
8
9   000006F0 00 00 00 00 00 00 00 00 00 00 00 00 00 00 00 00
10
11  00000700 01 01 03 02 80 84 1E 00 80 84 1E 00 E0 93 04 00
12
13  00000710 E0 93 04 00 E0 93 04 00 01 00 00 00 41 FA 41 FA
14
15
16  00000720 55
17
18  setpib -v <PIB_FILE> 704 data 00 12 7A 00 #modify data at
      offset 704 after converting to network-byte order
19
20  Result:
21  000006E0            00 00 00 00 00 00 00 00 00 00 00 00
22
23  000006F0 00 00 00 00 00 00 00 00 00 00 00 00 00 00 00 00
24
```

---

[3] The PIB values can be reverse-engineered if printed in hexadecimal format. For example, through trial and error of various offsets, we have managed to find the known values of TLL and modify them. Some vendors release the PIB structure too or enable modifications via their software tools. Hence, piggybacking the management messages can reveal the PIB offsets.

```
25  00000700 01 01 03 02 00 12 7A 00 80 84 1E 00 E0 93 04 00
26
27  00000710 E0 93 04 00 E0 93 04 00 01 00 00 00 41 FA 41 FA
28
29  00000720 55 55 55 55 55 55 55 55
30
31  plctool -i <ETH_INTERFACE> local -P <PIB_FILE> #send
       modified pib to device
```

Modifying the TTL with the PIB file can be different over different chipsets. We now give the commands for changing the TTL with INT6xxx chipsets. First, we read the PIB file as before (line 1). Second (line 2), we modify the PIB file with the xml2pib command that requires a .xml file (shown at line 7) for passing the offsets and modification instructions. Third (line 3), we verify the modification by a command that dumps the hex digits. Finally (line 4), we write the PIB file to the PLC interface as before:

```
1  int6k -i <ETH_INTERFACE> local -p <PIB_FILE> #read pib
2  xml2pib -f change_ttl.xml <PIB_FILE> #modify pib by using
      an .xml file
3  setpib -v <PIB_FILE> 0210 skip 64 #verify the modification
4  int6k -i <ETH_INTERFACE> local -P <PIB_FILE> -C 2 #write
      modified pib to device
5
6  change_ttl.xml:
7  <pib xmlns:xsi="http://www.w3.org/2001/XMLSchema-instance"
      xsi:noNamespaceSchemaLocation="piboffset.xsd"> <object
      name="PriorityTTL"> <offset>0210</offset> <length>16</
      length> <array> <dataHex>33333333</dataHex> <dataHex>a
      0860100</dataHex> <dataHex>e0930400</dataHex> <dataHex>e
      0930400</dataHex> </array> </object> </pib>
```

For future PIB files, one could read the hexadecimal form of the PIB file and search for the default TTL values to determine the correct offset. The default values are shown above. We have used this technique to reverse-engineer newer PIB versions and modify the TTL. The same procedure can be followed for other PIB parameters.

4.5   Network Configuration

Network configuration is extremely important for both residential and enterprise environments. In both environments, security and full coverage should be guaranteed. In this section, we show how to set up a PLC network.

As mentioned in Section 3.7, each PLC station must be authenticated by a CCo. If the station does not sense any CCo transmitting beacons, then it forms its own network. To avoid isolated PLC stations or modifications in the network structure, the network administrator can manually set the CCo. We found that this is particularly important in large PLC networks, such as our enterprise network that consists of twenty-five stations. In our testbed, we build two different PLC networks. These networks have different membership keys, which determine the logical network and the network encryption key (NEK). Thus only stations belonging to the same network can communicate with each other.

4.5.1 Setting the CCo

The station can be statically configured to always, never, or automatically be a CCo. The PLC devices are in an automatic (Auto) mode by default; in this mode, the station with the best channel conditions dynamically becomes the CCo. In small PLC networks such as residential ones, all stations are expected to have similar channel qualities, and the CCo setting is somewhat less crucial than in enterprise networks. In large-scale scenarios, the channel dynamics and station locations might require a guaranteed topology and CCo to avoid disruptions in communication. The station's CCo mode can be changed in the PIB file. First, we retrieve the PIB file from the PLC interface. Second, we modify the CCo mode by using `modpib -C`. Finally, we upload the modified PIB file back to the interface.

The commands for modifying the CCo mode of a station for INT6xxx chipsets are the following:

```
1  int6k -i <ETH_INTERFACE> -p <PIB_FILE> <MAC_ADDRESS>  #
       Retrieve the PIB file from the PLC interface
2  modpib -C <MODE> -v <PIB_FILE> # Modify CCo mode of <PIB_
       FILE>, <MODE>: 0 Auto, 1 Never, 2 Always CCo
3  int6k -i <ETH_INTERFACE> -P <PIB_FILE> <MAC_ADDRESS> #
       Update the PIB file of the PLC interface
```

Similarly, the commands for modifying the CCo mode of a station for QCA7xxx chipsets are the following:

```
1  plctool -i <ETH_INTERFACE> local -p <PIB_FILE>  # Retrieve
       the PIB file from the PLC interface
2  modpib -C <MODE> -v <PIB_FILE> # Modify CCo mode of <PIB_
       FILE>, <MODE>: 0 Auto, 1 Never, 2 Always CCo
3  plctool local -i <ETH_INTERFACE> -P <PIB_FILE> # Update the
       PIB file of the PLC interface
```

4.5.2 Setting the Membership Key

To establish a PLC network, we need to set up a security key (network membership key or NMK). The NMK is a 16-byte hexadecimal string used to encrypt the data on the network using 128-bit AES (see Section 7.2.1). By default, all PLC commercial devices come with the same default NMK, which introduces vulnerability to security attacks (security issues are discussed further in Chapter 7). We recommend modifying the NMK of any new PLC network.

The NMK of a PLC network can be modified by either commercial vendor tools or open source tools. Here we explain how to configure the NMK with open source Linux tools. We modify the PIB files of the interfaces by using Open PLC Utils. To change the NMK of a station with an INT6xxx chipset and with MAC address <MAC_ADDRESS>, we can execute the commands

```
1  int6k -p <PIB_FILE> <MAC_ADDRESS>  # Retrieve the PIB file
     from the PLC interface
2  modpib -N <NMK> -v <PIB_FILE> # Modify NMK
3  int6k -P <PIB_FILE> <MAC_ADDRESS> # Update the PIB file of
     the PLC interface
```

Similarly, for QCA7xxx PLC stations, we can run

```
1  plctool local -i <ETH_INTERFACE> -p <PIB_FILE>  # Retrieve
     the PIB file from the PLC interface
2  modpib -N <NMK> -v <PIB_FILE> # Modify NMK
3  plctool local -i <ETH_INTERFACE> -P <PIB_FILE> # Update the
     PIB file of the PLC interface
```

The NMK can be also set by using the int6k or amptool commands with the option -J. For instance, we can set the NMK by running the commands

```
1      int6k -J <NMK> # Modify NMK
2      amptool -J <NMK> # Modify NMK
```

In addition to modifying the NMK, a new station can join the PLC network by the so-called push-button. The push-button operation exists in commercial PLC devices and enables a manual level of security by pairing the stations. The push-button functionality can be also achieved by using the int6k or amptool commands with the option -B. This option has further actions called "join," "leave," "status," or "reset." In the join mode, the device will attempt to join an existing HomePlug AV network. In the leave mode, the device randomizes its HomePlug AV membership key, thus leaving the network. In the status mode, the command returns the authentication status, and finally, in the reset action, it restores to factory defaults. The LED indications on PLC devices indicate the push-button mode.

4.5.3　Retrieving Network Structure

After configuring the network structure, it is crucial to verify that the network works as expected. Various management messages retrieve the complete network structure and the station's role.

For example, by running faifa and selecting the 0x0014 MMType, we can get all stations with their TEI, MAC address, and CCo capabilities (whether the station can be selected as CCo):

```
1  Central Coordination Discover List Request
2  Choose the frame type (Ctrl-C to exit): 0x0014
3  Frame: Central Coordination Discover List Request (0x0014)
4
5  Dump:
6  Frame: Central Coordination Discover List Confirm (0015),
     HomePlug-AV Version: 1.1
7  Number of Stations: 11
```

```
 8   MAC address: 28:C6:8E:B6:E2:3E
 9   TEI: 1
10   Same network: Yes
11   SNID: 5
12   CCo caps: 00
```

By running `faifa` and selecting the `0x6038` MMType, we can retrieve information about the PLC network to which the station belongs (such as the NID), the station's role, the station's TEI, the network type, and the number of neighbors:

```
 1   Get Network Infos Request
 2   Choose the frame type (Ctrl-C to exit): 0x6038
 3   Frame: Get Network Infos Request (0x6038)
 4
 5   Dump:
 6   Frame: Get Network Infos Confirm (6039), HomePlug-AV
        Version: 1.1
 7   NID: C4 5B D7 10 80 1D 0D
 8   TEI: 0x0C (12)
 9   STA Role: 0x00 (Station)
10   MAC address: 28:C6:8E:B6:E2:3E
11   Access: 0x00 (In-Home network)
12   Number of neighbors: 0
```

The above commands print the PLC firmware MAC addresses of the stations. These MAC addresses will later be used for retrieving PHY/MAC statistics for PLC links in Sections 4.6 and 4.7. It is important to remember that we need the PLC firmware and not the Ethernet MAC addresses to obtain such PHY/MAC statistics.

As we note in Section 4.2, the PLC chip is bridged with an Ethernet interface.[4] To retrieve the list of Ethernet bridged stations, we run `faifa` and select the `0x6020` MMType as follows:

```
 1   Get Bridge Infos Request
 2   Choose the frame type (Ctrl-C to exit): 0x6020
 3   Frame: Get Bridge Infos Request (0x6020)
 4
 5   Dump:
 6   Frame: Get Bridge Infos Confirm (6021), HomePlug-AV Version
        : 1.1
 7   Bridging: Yes
 8   Bridge TEI: 0c
 9   Number of stations: 2
10   Bridged station 0 00:00:00:00:00:00
11   Bridged station 1 00:0D:B9:3E:A0:6E
```

Similar functionalities can be achieved with Open PLC Utils:

```
1        int6k -i <ETH_INTERFACE> -m  # for INT6xxx chips
2        amp -i <ETH_INTERFACE> -m  # for ARxxx chips
3
```

---

[4] Depending on the number of Ethernet ports of the device, the PLC chip can bridge more than one Ethernet interface.

```
4    Sample output:
5    network->NID = B0:F2:E6:95:66:6B:03
6    network->SNID = 14
7    network->TEI = 1
8    network->ROLE = 0x02 (CCO)
9    network->CCO_DA = 00:50:C2:A2:70:00
10   network->CCO_TEI = 1
11   network->STATIONS = 13
12
13   station->MAC = 00:50:C2:A2:70:1B
14   station->TEI = 7
15   station->BDA = 00:50:C2:A2:70:7C
16   station->AvgPHYDR_TX = 101 mbps
17   station->AvgPHYDR_RX = 104 mbps
18
19   station->MAC = 00:50:C2:A2:70:25
20   station->TEI = 8
21   station->BDA = 00:50:C2:A2:70:85
22   station->AvgPHYDR_TX = 036 mbps
23   station->AvgPHYDR_RX = 050 mbps
```

In lines 19 and 21 above, we observe the different MAC addresses shown in Figure 4.3.

4.6 PHY-Layer Measurements

PHY layer statistics are important for high quality-of-service and for understanding and troubleshooting the performance of individual PLC links. We discuss the PHY-layer metrics that are currently available on commercial devices.

4.6.1 Measuring Capacity

The station can measure the capacity to any other peer in the PLC network. A necessary step before requesting the PHY capacity from the PLC firmware is to initiate communication between the station and its peer. As explained in Section 2.7, the modulation scheme in PLC is referred to as a tone map, and it is unique per link. The tone maps are estimated and established only if there is data to be transmitted on the link. Thus, for capacity estimation, the user should first send some data, for example, by using ping, and then run faifa with MMType 0x6048. We discuss the precision of capacity estimation and its overheads in Section 5.5.

```
1    Get Network Stats Request
2    Choose the frame type (Ctrl-C to exit): 0x6048
3    Frame: Get Network Stats Request (0x6048)
4
5    Dump:
6    Frame: Get Network Stats Confirm (6049), HomePlug-AV
         Version: 1.1
7    MAC address: 28:C6:8E:B6:E2:3E
8    Average data rate from STA to DA: 9
9    Average data rate from DA to STA: 9
```

```
10   MAC address: 00:0B:3B:79:2D:53
11   Average data rate from STA to DA: 9
12   Average data rate from DA to STA: 9
13   MAC address: 00:0B:3B:79:2D:E7
14   Average data rate from STA to DA: 9
15   Average data rate from DA to STA: 9
```

The rates returned here are the BLE, as described in Section 2.7. Note that the above command returns the BLE for reception and transmission to all the peer stations in the network, identified by their internal PLC firmware MAC address. It returns 9 Mpbs for all links, as we have not exchanged any data with our peer stations, and 9 Mbps is simply a broadcast, default rate for PLC.

Alternatively, the user can use Open PLC Utils:

```
1   int6krate -i <ETH_INTERFACE> # for INT6xxx chips
2   amprate -i <ETH_INTERFACE> # for QCAxxx chips
3
4   Example output with both commands:
5   eth2 28:C6:8E:B6:E6:5E 00:0B:3B:79:2D:E7 TX 123 mbps
6   eth2 28:C6:8E:B6:E6:5E 00:0B:3B:79:2D:E7 RX 115 mbps
7   eth2 28:C6:8E:B6:E6:5E 00:0B:3B:79:36:FD TX 009 mbps
8   eth2 28:C6:8E:B6:E6:5E 00:0B:3B:79:36:FD RX 009 mbps
```

Lines 4 and 5 above show the PHY rates on the two directions of the link between the local station and the station with PLC firmware MAC address 28:C6:8E:B6:E6:5E. The above commands enable the option -u, which returns the *uncoded* rated (before FEC coding for errors). Thus, with these commands, we can easily compute the FEC rate used by the links. FEC rate is an indicator of link quality, as the higher it is, the better the quality is.

4.6.2   Measuring OFDM Modulation Statistics

The capacity-estimation procedure described in the previous section includes the PHY rate (BLE) and does not report the modulation per carrier. To get an estimated frequency response of the PLC channel and the modulation per carrier, open source tools provide additional functionalities. The user can retrieve the modulation per carrier and per tone map slot (see Section 2.7) for any peer with faifa and MMType 0x0A70.

```
1    Choose the frame type (Ctrl-C to exit): 0xa070
2    Frame: Get Tone Map Characteristics Request (0xA070)
3    Address of peer node?
4    00:0B:3B:79:2D:E7
5    Tone map slot?
6    0 -> slot 0
7    1 -> slot 1 ...
8    0
9
10   Dump:
11   Frame: Get Tone Map Characteristics Confirm (A071),
         HomePlug-AV Version: 1.0
12   Status: Success
13   Tone map slot: 00
```

```
14   Number of tone map: 06
15   Tone map number of activer carriers: 917
16   Modulation for carrier: 0 : QAM-16
17   Modulation for carrier: 1 : QAM-16
18   Modulation for carrier: 2: QAM-16
19   ................................................
20   Modulation for carrier: 130 : QAM-1024
21   Modulation for carrier: 130 : QAM-256
22   Modulation for carrier: 131 : QAM-64
23   Modulation for carrier: 131 : QAM-64
24   Modulation for carrier: 132 : QAM-64
25   Modulation for carrier: 132 : QAM-64
26   Modulation for carrier: 133 : QAM-16
27   Modulation for carrier: 133 : QAM-64
28   Modulation for carrier: 134 : QAM-64
29   Modulation for carrier: 134 : QAM-64
30   Modulation for carrier: 135 : QAM-64
31   Modulation for carrier: 135 : QAM-256
32   Modulation for carrier: 136 : QAM-256
33   Modulation for carrier: 136 : QAM-256
34   Modulation for carrier: 137 : QAM-1024
35   Modulation for carrier: 137 : QAM-1024
36   Modulation for carrier: 138 : QAM-1024
37   Modulation for carrier: 138 : QAM-1024
38   Modulation for carrier: 139 : QAM-1024
39   Modulation for carrier: 139 : QAM-256
40   Modulation for carrier: 140 : QAM-256
```

The user needs to enter the MAC address of the peer node (PLC firmware MAC address) and the tone map slot number, which is an integer between 0 and 6. The above `faifa` command returns the tone maps for reception from the peer node. The tone maps for reception and transmission of any link can be different due to asymmetry in PLC links. To retrieve the tone maps for transmission to the peer node, `Open PLC Utils` provide the `amptone` or `plctone` commands:

```
1   plctone -i <ETH_INTERFACE> local 00:0B:3B:79:2D:E7
```

These commands work similarly to `faifa`, but they return more statistics, such as SNR values per carrier. The `plctone` command works only with QCA7xxx chips.

4.6.3   Measuring PHY Errors

An important PHY-layer metric is the PB error rate (PB_{err}). High error rates can lead to high access delays to the medium as well as low effective throughput. PLC devices provide a lot of information about PHY errors and MAC collisions for both transmission to and reception from a peer station. The user can retrieve these statistics with `faifa` and MMType `0xA030`:

```
1   Get Link Statistics Request
2   Choose the frame type (Ctrl-C to exit): 0xA030
3   Frame: Get Link Statistics Request (0xA030)
4   Direction ?
5   0: TX
6   1: RX
```

```
 7   2: TX and RX
 8   2
 9   Link ID ?
10   1
11   Address of peer node?
12   00:0B:3B:79:2D:E7
13
14   Dump:
15   Frame: Get Link Statistics Confirm (A031), HomePlug-AV
         Version: 1.0
16   Status: Success
17   Link ID: 01
18   TEI: 03
19   Direction: Tx
20   MPDU acked.......................: 2
21   MPDU collisions..................: 0
22   MPDU failures....................: 0
23   PB transmitted successfully......: 2
24   PB transmitted unsuccessfully...: 0
25   Direction: Rx
26   MPDU acked.......................: 3
27   MPDU failures....................: 0
28   PB received successfully.........: 3
29   PB received unsuccessfully.......: 0
30   Turbo Bit Errors passed..........: 10
31   Turbo Bit Errors failed..........: 0
32   -- Rx interval 0 --
33   Rx PHY rate......................: 00
34   PB received successfully.........: 0
35   PB received failed...............: 0
36   TBE errors over successfully.....: 0
37   TBE errors over failed...........: 0
38   -- Rx interval 1 --
39   Rx PHY rate......................: 00
40   PB received successfully.........: 0
41   PB received failed...............: 0
42   TBE errors over successfully.....: 0
43   TBE errors over failed...........: 0
44   -- Rx interval 2 --
45   Rx PHY rate......................: 00
46   PB received successfully.........: 0
47   PB received failed...............: 0
48   TBE errors over successfully.....: 0
49   TBE errors over failed...........: 0
50   -- Rx interval 3 --
51   Rx PHY rate......................: 123
52   PB received successfully.........: 1
53   PB received failed...............: 0
54   TBE errors over successfully.....: 10
55   TBE errors over failed...........: 0
56   -- Rx interval 4 --
57   Rx PHY rate......................: 00
58   PB received successfully.........: 0
59   PB received failed...............: 0
60   TBE errors over successfully.....: 0
61   TBE errors over failed...........: 0
```

```
62  -- Rx interval 5 --
63  Rx PHY rate.....................: 00
64  PB received successfully........: 0
65  PB received failed..............: 0
66  TBE errors over successfully....: 0
67  TBE errors over failed..........: 0
```

To retrieve the above statistics, the user has to enter the PLC firmware MAC address of the peer station (see lines 11–12), the direction of the link (i.e., transmission, reception, or both; see lines 4–8), and the Link ID (see lines 9–10), which is the PLC priority of the link. The information obtained is rich: for transmission, we have the number of MPDUs collided and acknowledged (lines 20–22) and the PBs successfully or unsuccessfully transmitted (lines 23–24). For reception, we have similar statistics and the PHY rate (BLE) for each tone map slot called "Rx interval" above (see, e.g., line 32).

Similarly, the user can retrieve the above statistics with Open PLC Utils commands, such as int6kstat, int6krate, or ampstat. For example, to get PHY rates, we can run

```
1  int6krate -i <ETH_INTERFACE>
```

To retrieve PHY errors, we can execute

```
1    int6kstat -i <ETH_INTERFACE> -d <LINK_DIRECTION> -s <
     LINK_ID> -p <PEER_NODE_MAC_ADDRESS>
```

For certain experimental studies and for isolating network malfunctioning phenomena, it is useful to reset the above statistics to get fine-grained or periodic measurements. Open PLC Utils enables resetting the PB statistics by using the command ampstat with the option -C:

```
1    int6kstat -i <ETH_INTERFACE> -C -d <LINK_DIRECTION> -s <
     LINK_ID> -p <PEER_NODE_MAC_ADDRESS>
```

4.7   MAC-Layer Configuration and Measurements

MAC-layer measurements represent the level of contention in the network as well as communication overheads for control and management. In this section, we describe how to retrieve MAC-layer statistics both at long time intervals and on a per-MPDU basis.

4.7.1   Modification of Packet/Traffic Priority

In Section 3.4, we note that PLC employs a prioritized CSMA/CA method, where the high-priority classes have immediate access to the medium. This enables high QoS for delay-sensitive and best-effort traffic. To change the priority of the PLC frames and to conduct tests with various traffic classes, we modify the type-of-service (ToS) field of the IP layer header or the VLAN priority of the Ethernet header. For instance, to modify the PLC priority to CA3, we set the ToS field of the IP header to a 192 value. We have achieved this with running `iperf` or generating our own traffic. Commodity devices often allow the user to manually assign certain traffic types to CA0–CA3 priorities. To identify and prioritize network traffic, the user has to use commodity software tools. Service-based QoS follows the IEEE 802.1p Layer 2 QoS Protocol for traffic prioritization, and devices support ToS or VLAN prioritization. Traffic prioritization can also depend on Ethernet source/destination addresses, IP source/destination addresses, and TCP source/destination ports. Vendor software can provide interfaces for configuring the application to ToS mapping.

4.7.2   Measuring Collisions

PLC employs a CSMA/CA mechanism that can guarantee low but nonzero collision probability between different stations (see Section 3.5). Collisions indicate the level of contention in the network and the presence of hidden stations. Collision rate increases with the number of transmitting stations and usually explodes when there exists a hidden terminal. To measure collision probability, we use MMType 0xA030 (Get Link Statistics Request) with the two open source tools that we have presented. The command `ampstat` can reset to 0 or retrieve the number of acknowledged and collided MPDUs given the peer-station MAC address, the priority, and the flow direction (transmission or reception), as we have discussed in Section 4.6.3. The commands are the following:

```
 1   ampstat -i <ETH_INTERFACE> -C -d <LINK_DIRECTION> -s <
     LINK_ID> -p <PEER_NODE_MAC_ADDRESS > # reset
     statistics at beginning of test
 2   ........ # run test
 3   ampstat -i <ETH_INTERFACE>  -d <LINK_DIRECTION> -s <
     LINK_ID> -v -p <PEER_NODE_MAC_ADDRESS > # measure
     statistics , with verbose mode that prints byte
     contents of the 0xA031 message

 4
 5
 6   Sample output:
 7   00000000 00 B0 52 00 00 01 00 50 C2 A2 70 75 88 E1 00
     30 ..R....P..pu...0
 8   00000010 A0 00 B0 52 00 00 01 00 50 C2 A2 70 12 00 00
     00 ...R....P..p....
 9   00000020 00 00 00 00 00 00 00 00 00 00 00 00 00 00 00
     00 ...............
10   00000030 00 00 00 00 00 00 00 00 00 00 00 00
                 ............

11
12   00000000 00 50 C2 A2 70 75 00 50 C2 A2 70 13 88 E1 00
     31 .P..pu.P..p....1
13   00000010 A0 00 B0 52 00 00 01 01 5B 3C 00 00 00 00 00
     00 ...R....\[<......
14   00000020 FA 27 00 00 00 00 00 00 00 00 00 00 00 00 00
     00 .'..............
15   00000030 24 F8 07 00 00 00 00 00 F8 0D 00 00 00 00 00
     00 ...............
16   00000040 00 00 00 00 00 00 00 00 00 00 00 00 00 00 00
     00 ...............
```

```
17                  . . . . . . . . . . . . . . . . . . . . . . .
18      00000120 00  00  00  00  00  00  00  00  00  00  00  00  00  00  00
        00 . . . . . . . . . . . . . . .
19      00000130 00  00  00  00  00  00  00
20
21      DIR ----------- PBs PASS ----------- PBs FAIL PBs ERR
           ---------- MPDU ACKD ---------- MPDU FAIL
22      TX                522276                3576    0.68%
                     15451                          0    0.00%
```

As with every management message presented and used in this chapter, we can get the above MM structure from the code of open source tools. We request the number of collided frames with MMType 0xA030. From the response with MMType 0xA031, we use the bytes 25–32 of this response, which represent the number of acknowledged frames (including collided ones), and the bytes 33–40, which represent the number of collided frames. Let C_i be the number of collided frames transmitted by station i, and let A_i be the number of acknowledged frames transmitted by station i. To evaluate the collision probability in the network, we compute $\sum_{i=1}^{N} C_i / \sum_{i=1}^{N} A_i$. Observe that we divide only by the sum of the acknowledged frames, excluding the collided frames from the denominator, as the 1901 standard allows selective acknowledgments for all the physical blocks contained in a frame [5]. When a collision occurs and the preambles of the collided frames could be decoded (due to the robust modulation with which they are transmitted), the destination acknowledges the frame, with an indication that all the physical blocks are received with errors, which yields a collision. Recall that stations can transmit multiple MPDUs through a single CSMA/CA access (see Section 3.3). Up to four MPDUs may be transmitted in a burst. While this number indicates the upper limit, the actual number of MPDUs per burst supported by a station depends on channel conditions and station capabilities [7]. In IEEE 1901, bursts contend for the medium and not individual MPDUs; the method presented above measures MPDU collision probabilities, and not the collision probability per channel access. However, in our testbed, we find that the stations in saturated experiments almost always use bursts with two MPDUs (this was measured by using the SoF). This means that most bursts contain two MPDUs and that we can approximate the collision probability per channel access by the MPDU collision probability. Thus, to evaluate the collision probability per channel access via testbed measurements, we use the number of collided and acknowledged MPDUs, because this is the only information we can obtain from the INT6xxx PLC chips and a very good approximate of the collision probability per channel access.

4.7.3 Capturing MPDUs

Commercial devices enable capturing the PLC header of every MPDU transmitted over the channel. To this end, the station enters a promiscuous mode called "sniffer."[5] To enable the sniffer mode, either faifa with MMType 0xA036 or ampsnif from Open PLC Utils can be used. A PLC device with sniffer mode enabled can retrieve all MPDU

---

[5] We recommend enabling this mode only for experiments, as it consumes CPU resources and could harm the actual communication or limit the router's capabilities.

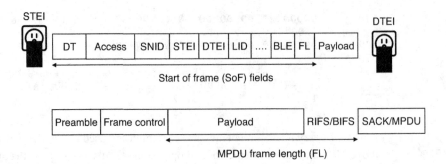

Figure 4.5 The most important fields of the frame control of PLC MPDUs. The computation of the field frame length (FL).

frame controls through any packet analyzer, such as `tcpdump` or `wireshark`, and can be processed in userspace by tools such as `Click!` modular router.

The PLC header or SoF delimiter can provide a lot of information, the most useful being shown in Table 4.5 and Figure 4.5. The user can retrieve the arrival timestamp of the PLC frame, which is given in reference to a 25 MHz PLC clock.[6] To compute the actual timestamp or difference between timestamps of consecutive frames, we need to divide the reported timestamp by 25. The "delimiter type" field gives us the type of the MPDU, that is, frame, beacon, ACK, RTC/CTS, and sounding. The PLC frame header gives the BLE and the tone map index with which each frame is modulated and coded, which enables studying the real-time variation of BLE and its periodicity with respect to the AC cycle. The BLE is encoded as required by the standard [5]: the first 5 bits are the mantissa (*mant*) and the last 3 bits the exponent (*exp*) in the formula $BLE = (32 + mant) \times 2^{exp-4} + 2^{exp-5}$. Hence, to parse this field, the user can run the following commands in Python:

```
1   integer = int(BLE_from_frame,16)
2   string = '{0:08b}'.format(integer)
3   exp = string[5:8]
4   mant = string[0:5]
5   # formula given by 1901 standard
6   BLE = (32+int(mant,2))*2**(int(exp,2)-4)+2**(int(exp,2)-5)
```

The priority (or link ID) and the source/destination TEIs of the frame are also included in the PLC frame header. Finally, burst information is given in order for the station to acknowledge the MPDUs. The frame duration indicates the duration of the MPDU in multiples of 1.28 μs [5], hence the user needs to multiply the value of this field with 1.28 to get the actual duration. The value of this duration includes the subsequent interframe space, as shown in Figure 4.5. The number of MPDUs remaining for the current burst equals to 0 when the last MPDU in the burst is transmitted. The selective acknowledgment of PLC follows this last MPDU.

The following output shows the activation of sniffer mode with `faifa` as well as a

---

[6] According to IEEE 1901 standard, each device should be synced using a 25 MHz clock.

Table 4.5 Metrics and measurement methods

Metric	Notation
Arrival timestamp	t
Delimiter Type (DT): Frame, beacon, ACK, RTC/CTS, Sound	-
Bit-loading estimate of frame	BLE
Tone map index	TI
Priority of the frame	$LinkID$
Source station of the frame	$STEI$
Destination station of the frame	$DTEI$
Frame duration	FL
Number of remaining MPDUs for current burst	MPDUCnt

Note. For more information about the SoF delimiter fields, see Section 3.2.1 or the IEEE 1901 standard [5].

dump example of a PLC delimiter that is a beacon (delimiter type is 0). Notice that faifa prints a lot of information in hexadecimal form:

```
 1  Choose the frame type (Ctrl-C to exit): 0xa034
 2  Frame: Sniffer Mode Request (0xA034)
 3  Sniffer mode?
 4  0: disable
 5  1: enable
 6  2: no change
 7
 8  Dump:
 9  Frame: Sniffer Mode Indicate (A036), HomePlug-AV Version:
       1.0
10  Type: Regular
11  Direction: Rx
12  System time: 162945414122258
13  Beacon time: 3148
14  Delimiter type: 0
15  Access: No
16  SNID: 5
17  STEI: 35
18  DTEI: 7f
19  Link ID: 6d
20  Contention free session: No
21  Beacon detect flag: No
22  HPAV version 1.0: Yes
23  HPAV version 1.1: Yes
24  EKS: 5
25  Pending PHY blocks: 60
26  Bit loading estimate: 02
27  PHY block size: No
28  Number of symbols: 1
29  Tonemap index: a
30  HPAV frame length: a02
31  MPDU count: 0
32  Burst count: 1
33  Convergence layer SAP type: 2
34  Reverse Grant length: 20
35  Management MAC Frame Stream Command: 4
```

```
36  Data MAC Frame Stream Command: 4
37  Request SACK Retransmission: No
38  Multicast: No
39  Different CP PHY Clock: No
40  Multinetwork Broadcast: No
41  Frame control check sequence: 914b47
42  Delimiter type: 4
43  Access: No
44  SNID: c
45  Beacon timestamp: 2148587355 (0x8010d75b)
46  Beacon transmission offset 0: 36125 (0x8d1d)
47  Beacon transmission offset 1: 1 (0x0001)
48  Beacon transmission offset 2: 1 (0x0001)
49  Beacon transmission offset 3: 1282 (0x0502)
50  Frame control check sequence:  1 6 0
```

An example of MPDU that carries data is the following:

```
 1  Direction: Rx
 2  System time: 26657900886
 3  Delimiter type: 1
 4  STEI: 03
 5  DTEI: 06
 6  Link ID: 01
 7  HPAV version 1.0: No
 8  HPAV version 1.1: No
 9  Bit loading estimate: 26
10  HPAV frame length: 530
11  MPDU count: 1
```

Note that the data above is transmitted at priority CA1 (Link ID at line 6 is 01) from TEI 03 to TEI 06 (lines 4 and 5). Information about BLE, the MPDU duration, and the location of the MPDU in the burst is provided in lines 9, 10, and 11, respectively.

faifa always prints the captured header as if it results from a data frame: the same information is printed independently of the delimiter type of the MPDU header. The printed information needs to be updated to expose the correct information for beacons, ACKs, RTS/CTS, and sounding delimiter types. Thus the user should take into account the above information only for "Delimiter type: 1" that represents data frames. The contention-free session printed in line 20 above indicates whether there was a TDMA transmission. For further information, one can find the structure for the rest of the delimiter types in the IEEE 1901 standard [5]. We next discuss useful measurement applications based on MPDU capture.

4.7.4   Measuring MPDUs per Priority Class

The tools enable to capture the number of frame delimiters for different priority classes. Data can be of class CA0–3, whereas MMs are transmitted at priority class CA2 or CA3 over the medium. PLC MMs are not forwarded to the Ethernet interface, which means that they are not received by the user. By capturing the PLC frame header, the sniffer mode enables the user to capture their presence and potential overheads. In order to isolate them, the user can send traffic at lower priority, such as CA0 and CA1. With this configuration, only MMs are tagged with priority CA2 or higher, and their frame header can be distinguished by the user. Unfortunately, only the frame header of MMs can be currently captured. Therefore we can measure their overhead with respect to data frames, their frequency, and their size. We present results of this study in Chapter 6. As an example, we can measure the MMs transmitted by looking at priority class LINK_ID to 2 or 3, with faifa sniffer mode or with the following commands:

```
1    ampstat -i <ETH_INTERFACE> -C -d <LINK_DIRECTION> -s <
     LINK_ID> -p <PEER_NODE_MAC_ADDRESS> # reset
     statistics at beginning of test
2    ........ # run test
3    ampstat -i <ETH_INTERFACE>  -d <LINK_DIRECTION> -s <
     LINK_ID> -v -p <PEER_NODE_MAC_ADDRESS> # measure
     statistics
```

Similarly, in tests with multiple priority classes contending, we can measure the proportion of SoF per class with the commands above.

4.7.5 Measuring Short-Term Statistics

We have used the MPDU sniffer mode to capture short-term fairness statistics on PLC (the results are presented in Chapter 6). This is a case where faifa is extremely useful, since simply capturing Ethernet packets does not work with PLC. Owing to frame aggregation with PLC, we cannot employ tcpdump to capture the frames, as this tool only captures the Ethernet packets. The number of Ethernet packets contained in a PLC frame is variable, hence identifying the sequences and interarrival times of actual PLC frames is challenging. To address this challenge, we use faifa to capture the SoF delimiters of PLC frames and to retrieve useful metrics, such as those introduced in Table 4.5. To study fairness, we need to know the station identifications of all transmitted PLC frames. Faifa prints all the fields of SoF, including the source TEI, the MPDU count, and the arrival time of the frame, which are of interest in this experiment.

As an example, to study fairness between contending stations, we can use the following fields as printed by faifa:

```
1  Direction: Rx
2  System time: 26657900886
3  Delimiter type: 1
4  STEI: 03
5  DTEI: 06
6  Link ID: 01
7  MPDU count: 1
```

We need to make sure that the direction of transmission is correct (Rx if we are capturing packets at a station not transmitting during the experiment), that the delimiter contains a frame (Delimiter type: 1 at line 3), that the priority of the packet is correct (Link ID: 01 [CA1] at line 6, as set in our experiment), and that we count each burst once by using MPDU count field. Recall that bursts of MPDUs contend for the medium, hence to study the fairness of CSMA/CA access, we need to count frames with MPDU count: 1 and MPDU count: 0 as a single access to the medium. Finally, for such a study, we need to record the system time and the STEI and analyze the sequence of burst transmissions by different STEIs.

4.8   Building Your Own PLC Tools with `Click`!

The previous sections present the most important PHY/MAC metrics that we can measure with commodity PLC devices. In addition, we show how to configure single devices and network topology. We have used open source tools that work with the most popular vendors. However, there might be cases where running or installing such tools is not simple, causes overheads, or creates compatibility issues with other vendors. For example, in case of hybrid networks using Wi-Fi and PLC, we might avoid calling userspace programs that cause high latency, and we may choose to request and receive these statistics in a driver or kernel module. Typically, we would need to retrieve PHY/-MAC statistics for our own routing or path estimation modules over hybrid networks. In this section, we show how to program PLC interactions in any language, and we use as an example the C++ language and the `Click`! router. We will also use `Click`! later in Chapter 8, where we work on hybrid networks and implement a 2.5 layer (a layer between IP and MAC, compliant with IEEE 1905) hybrid PLC/Wi-Fi architecture from scratch. The code used in this chapter is uploaded on the public Github folder PLC Click Elements [10].

4.8.1   `Click`! Router Primer

`Click`! can be compiled as userspace or kernel module. A networking device driver can receive and send packets. The kernel module interacts with the device drivers, whereas the userspace module uses packet sockets (on Linux).

`Click`! provides great flexibility in packet processing and has a great number of open source functions that the community has contributed. Although it is called a router and can configure IP layer and routing, `Click`! processes packets at lower (MAC) or higher layers (application). For instance, it can send or reply to ARP (address resolution protocol) packets[7] or can manipulate the Ethernet header. The packets flow through a directed graph of configuration or processing modules, called elements. An element, which is a vertex on the graph, can have simple processing tasks, such as classifying the packet and forwarding to appropriate output, or complex tasks, such as scheduling [11]. The directed graph and relationship between elements are refined in a script with extension `.click`.

Figure 4.6 presents the main components of our script `plc_elem.click`.[8] The element **FromDevice** receives packets from a networking device. In our example, we have set this device to `eth2`, which is the Ethernet device that bridges the PLC chip. The element **Host** receives packets destined to this machine and delivers them to the kernel. It is also the element that receives packets from the kernel. An element can have any number

---

[7] ARP translates MAC addresses to IP addresses.
[8] See `https://github.com/christinavl/plc-click-elements/blob/master/plc_elem.click`.

of inputs and outputs, which can be predefined while programming the element or while declaring a new element in a `Click!` script. In case of multiple inputs, the input number is depicted at the head of the arrows in Figure 4.6. Similarly, outputs are depicted at the beginning of the arrow (if the element has multiple outputs, otherwise omitted). In addition, the element can have configuration strings or arguments. Any element in a script can be reused and referenced in `Click!`'s graph if declared by a name: for instance, in Figure 4.6, we have defined **arpr::ARPResponder** and **arpq::ARPQuerier**, which are reused as **arpr, arpq**. The rest of the elements cannot be reused in the script. Connections between elements are expressed with arrows ->. An example of `Click!` declaration and connection between classes is the following:

```
1     /////////////////////// From and To eth2
      /////////////////////////////////////
2     // Classifier from input eth2
3     // 0. ARP queries
4     // 1. ARP replies
5     // 2. IP
6     // 3. Other
7     cl_in :: Classifier(12/0806 20/0001, 12/0806 20/0002,
         12/0800, -);

8
9     // Elements for handling ARP requests and replies
10    arpq :: ARPQuerier(eth2);
11    arpr :: ARPResponder(eth2);
12
13    sendQueue_eth :: Queue(200) -> ToDevice(eth2, DEBUG
         false);
14    // ARP queries need to be treated with a reply that is
         sent to eth2
15    cl_in[0] -> arpr -> sendQueue_eth;
16    // Deliver ARP replies to ARP querier.
17    cl_in[1] -> [1]arpq;
18    // Discard non-IP packets
19    cl_in[3] -> Discard();
```

We now describe the functions of each element in Figure 4.6.

FromDevice receives packets from a device driver (e.g., Ethernet or Wi-Fi). As these packets can be IP, ARP, or other EtherType, usually it is followed by a classifier. Packets are represented and forwarded between elements using pointers in C++, as we will see later.

PLCElement classifies the incoming packets as either the PLC MMs we need or other regular traffic. In case the incoming packets are PLC frames, then this element performs some operations (e.g., stores the incoming statistics), and if needed, it replies to the PLC frame by creating new packets and forwarding them to output "1," connected to a queue for interface eth2. In case of regular traffic, this element forwards the packets to another classifier that distinguishes ARP, IP, or other traffic.

Classifier determines whether the packet is an ARP request, ARP reply, IP, or other. Classifier uses the pattern "byte offset/value" to determine the EtherType of the packet.

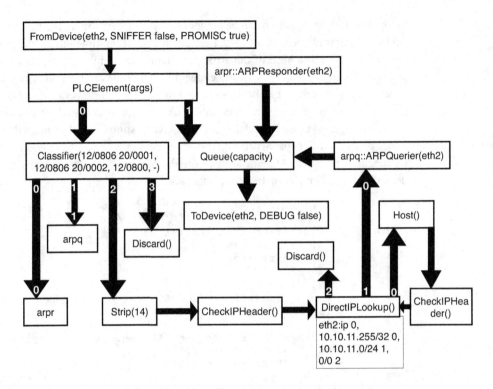

Figure 4.6 Our simple Click! design that incorporates a PLC element "PLCElement."

For instance, 12/0806 is an ARP packet, and with byte offset 20/0001 or 20/0002, we can classify the packet as ARP request and ARP reply. If the packet is ARP request, it is forwarded to **ARPResponder** which replies to the ARP request and forwards the reply to the appropriate device. If the packet is an ARP reply, it is forwarded to input "1" of **ARPQuerier**, which updates the ARP table with the IP/MAC addresses mapping included in ARP reply.

Strip(arg) simply removes arg bytes from the packet, which in this script is the MAC protocol header with 14 bytes.

CheckIPHeader makes sure that the incoming packet has a valid IP header and address. It emits valid packets in its output "0" and discards the rest.

DirectIPLookup is a simple routing element that receives IP packets from the device or the host. IP packets that belong to an IP of our machine have to be delivered to the host (kernel). In this example, we show the routing table below the element: we deliver the IP packets with IP destination address the one of eth2 interface (retrieved as eth2:ip in the Click! script) to the host, using output "0" of Routing. Packets to local network are delivered to **ARPQuerier** and then to **ToDevice** element for transmission (in this case, via the PLC channel.) **ARPQuerier** queues the packets if the ARP address of the

IP destination is not known and sends an ARP request for this IP address. The rest of the packets that arrive in **DirectIPLookup** are discarded, as we do not need them in this implementation.

4.8.2   Implementation of a `Click`! Element in C++ Language

To interact with the PLC chip, MMs have to be encapsulated in an Ethernet header with EtherType `0x88E1` (EtherType for Homeplug AV).[9] Our elements give simple examples on how to reuse the open source tools described in this chapter. `Click`! router can be used for testing and deploying hybrid networks, hence these elements are useful for network architects and researchers.

We first create a header file with definitions for protocols, MMs, and message headers. We name this file `PLCStats.h`.[10] After the MAC header (already defined by `Click`!), we define the Homeplug header with a `struct` as follows:

```
1  / Standard structure of the header of a HomePlug frame
2  // The standard messages start with MMType 0x60
3  // Vendor specific messages start with MMType 0xA0
4  struct click_hp_av_header {
5      uint8_t  version;
6      uint16_t MMType;
7  } CLICK_SIZE_PACKED_ATTRIBUTE;
8  //////////////////////////////////////////////
```

Then, we need to define the headers for every different MM defined by MMType. In our example, we use the vendor-specific MM that measures packet error statistics. For the error statistics MM, we define the following `Click`! header:

```
1  struct click_hp_av_error_stats_req {
2      uint8_t   oui[3];
3      uint8_t   control;
4      uint8_t   direction;
5      uint8_t   link_id;
6      uint8_t   macaddr[6];
7  } CLICK_SIZE_PACKED_ATTRIBUTE;
```

The "error statistics" MM starts with the 3 bytes of the vendor-specific OUI and 1 byte for control that is always set to 0 for the request MM. It is followed by 1 byte for the direction of the link, that is, transmission versus reception (0 or 1, respectively), as we ask for the error statistics per peer MAC address (last 6 bytes in our header above). In addition to the direction of the link, we can get error statistics per packet priority, hence we use the link_id byte in the header.

We also need to define the reply struct for every MMType. In our example, we have multiple structs for receive/transmit statistics and for the MM reply itself:

```
1      struct tx_link_stats {
2          uint64_t   mpdu_ack;
3          uint64_t   mpdu_coll;
4          uint64_t   mpdu_fail;
```

[9] EtherType is 0x887B for older HomePlug 1.0 released in June 2001.
[10] https://github.com/christinavl/plc-click-elements/blob/master/PLCStats.h.

```
5            uint64_t    pb_pass;
6            uint64_t    pb_fail;
7       } CLICK_SIZE_PACKED_ATTRIBUTE;
8
9       struct rx_link_stats {
10           uint64_t    mpdu_ack;
11           uint64_t    mpdu_fail;
12           uint64_t    pb_pass;
13           uint64_t    pb_fail;
14           uint64_t    tbe_pass;
15           uint64_t    tbe_fail;
16           uint8_t     num_rx_intervals;
17           struct rx_interval_stats    rx_interval_stats[0];
18       } CLICK_SIZE_PACKED_ATTRIBUTE;
19
20
21      struct click_hp_av_error_stats_rep {
22           uint8_t     oui[3];
23           uint8_t     mstatus;
24           uint8_t     direction;
25           uint8_t     link_id;
26           uint8_t     tei;
27           union {
28               tx_link_stats tx;
29               rx_link_stats rx;
30               struct {
31                   tx_link_stats txboth;
32                   rx_link_stats rxboth;
33               } CLICK_SIZE_PACKED_ATTRIBUTE;
34           } CLICK_SIZE_PACKED_ATTRIBUTE;
35      } CLICK_SIZE_PACKED_ATTRIBUTE;
```

Finally, we define the headers for both request and reply MMs:

```
1   #define ERROR_STATS_REQ    0x30a0
2   #define ERROR_STATS_REP    0x31a0
```

After defining the MMType for error statistics request and reply, we are ready to code our element. The same process can be repeated for multiple MMs, as we show in our folder in GitHub [10].

We now focus on the code for sending ERROR_STATS_REQ and receiving ERROR_STATS_REP, as defined above. Every Click! element is a C++ class, so we define the errorstatsreq.{cc,hh} files for our new PLC element. The header file is used to define the new element, its properties, and its public and private functions:

```
1   // File errorstatsreq.hh
2   #ifndef CLICK_ERRORSTATSREQ_HH
3   #define CLICK_ERRORSTATSREQ_HH
4   #include <click/element.hh>
5   #include <clicknet/ether.h>
6   #include <click/etheraddress.hh>
7   #include <click/sync.hh>
8   #include <click/timer.hh>
9   #include "PLCStats.h"
10
11  CLICK_DECLS
```

```
12
13   class ErrorStatsReq : public Element { public:
14
15       ErrorStatsReq();
16       ~ErrorStatsReq();
17
18       EtherAddress _src;
19       EtherAddress _dst;
20       int _prio;
21       int _dir;
22
23       const char *class_name() const  { return "ErrorStatsReq
           "; }
24       const char *port_count() const  { return "1/2"; }
25       const char *processing() const  { return PUSH; }
26       void *cast(const char *name);
27       void run_timer(Timer *);
28       int initialize(ErrorHandler *errh);
29       int configure(Vector<String> &, ErrorHandler *);
30       void push(int,Packet *);
31
32       private:
33       Timer _expire_timer_ms;
34
35       void sendErrorStatsReq();
36       void print_tx_stats(tx_link_stats *);
37       void print_rx_stats(rx_link_stats *);
38       void processErrorStatsRep(click_hp_av_error_stats_rep
           *);
39   };
40   CLICK_ENDDECLS
41   #endif
```

First, we have to include default header files of Click! (lines 2–9) for class Element, Ethernet header and addresses, Sync, and Timer, as we will send ERROR_STATS_REQ periodically and we need to include timer functions. Our element **ErrorStatsReq** extends the **Element** class defined in the router, which has a default constructor and destructor. When initialized, it uses a specific source and destination MAC address, which are variables _src and _dst. In addition, we have included the packet priority and direction as configuration arguments, declared with _prio, _dir. The class_name() function at line 23 returns the name of the element to be used by the Click!, and in our case, it is **ErrorStatsReq**. The port_count() at line 24 declares the number of inputs and outputs of the element in the format "inputs/outputs." Our element has one input and two outputs. An element can be defined as "PUSH/PULL" processing. Upstream elements initiate pull calls (such as the input ToDevice element), whereas downstream ones "push" packets to outputs. In a Click! script, pull outputs can only be connected to pull inputs, and the same holds for push outputs. **ErrorStatsReq** is a PUSH element, connected to the PUSH output of the **FromDevice** element. In our header file above, we declare all these element properties and in addition its public and private functions.

We now move to the main element functions in errorstatsreq.cc. First, we have to implement the configuration function such that the element reads our arguments in the Click! script:

```
1  // From our Click! scripts
2  ErrorStatsReq(SRC eth2:eth, DST 00:0D:B9:3D:C2:AA, PRIORITY
     1, DIRECTION 1)
```

The code below implements the configurations by using the Element configure() function and creating the mapping between the variable names in the Click! script and the ones in the C++ class. For example, the MAC address "SRC" is mapped to variable _src:

```
1  int
2  ErrorStatsReq::configure(Vector<String> &conf, ErrorHandler
     *errh)
3  {
4      return Args(conf, this, errh).read_m("SRC", _src)
5      .read_m("DST", _dst)
6      .read_m("DIRECTION", _dir)
7      .read_m("PRIORITY", _prio)
8      .complete();
9  }
```

The most important functionality of an element is its packet processing, implemented usually with the "push/pull" functions:

```
1  void
2  ErrorStatsReq::push(int port, Packet *p)
3  {
4      click_ether *e = (click_ether *) p->data();
5      if(e->ether_type == htons(ETHERTYPE_HP_AV)) {
6          click_hp_av_header *hpavh = (click_hp_av_header *)
             (e + 1);
7          if(ntohs(hpavh->MMType) == ERROR_STATS_REP) {
8              processErrorStatsRep((click_hp_av_error_stats_
                 rep*)(hpavh+1));
9              p->kill();
10         } else {
11             output(0).push(p);
12         }
13     } else {
14         output(0).push(p);
15     }
16 }
```

In our code, we first retrieve the Ethernet header of the incoming pointer of the packet (line 4). If the packet has the HP_AV EtherType, we proceed to further processing. Otherwise, we push it in the output 0 of our element (output(0).push(p); at line 14), as it might be an ARP or IP packet that needs different processing by other elements. Similarly, if the packet has HP_AV EtherType but not the appropriate MMType, we forward it to output 0 (line 11). We process only ERROR_STATS_REP in this element, with the function processErrorStatsRep(). After processing packet p, we discard the packet, as it is no longer needed or consumed by any other element. This is done by the command p->kill(); (line 9).

Processing the ERROR_STATS_REP can be used for many purposes, such as informing another element about PLC errors or printing information, which we implement here as an example:

```
1  void
2  ErrorStatsReq::processErrorStatsRep(click_hp_av_error_stats
   _rep *error_rep){
3      switch(error_rep->mstatus) {
4          case HPAV_SUC:
5              click_chatter("[ErrorStatsReq] Status of
                  received MME: Success\n");
6              break;
7          case HPAV_INV_CTL:
8              click_chatter("[ErrorStatsReq] Status of
                  received MME: Invalid control\n");
9              break;
10         case HPAV_INV_DIR:
11             click_chatter("[ErrorStatsReq] Status of
                  received MME: Invalid direction\n");
12             break;
13         case HPAV_INV_LID:
14             click_chatter("[ErrorStatsReq] Status of
                  received MME: Invalid Link ID\n");
15             break;
16         case HPAV_INV_MAC:
17             click_chatter("[ErrorStatsReq] Status of
                  received MME: Invalid MAC address\n");
18             break;
19     }
20
21     if (error_rep->direction == HPAV_SD_TX) {
22         print_tx_stats(&(error_rep->tx));
23     } else if  (error_rep->direction == HPAV_SD_RX) {
24         print_rx_stats(&(error_rep->rx));
25     } else if (error_rep->direction == HPAV_SD_BOTH) {
26         print_tx_stats(&(error_rep->txboth));
27         print_rx_stats(&(error_rep->rxboth));
28     } else {
29         click_chatter("[ErrorStatsReq] Unknown direction.")
               ;
30     }
31
32     return;
33 }
```

In the above code, we get as an input the `click_hp_av_ERROR_STATS_REP` header. We first print the status returned by the MM, using the existing `click_chatter` function. Then, we print the statistics contained in the MM, depending on the traffic direction (HPAV_SD_TX, HPAV_SD_RX, HPAV_SD_BOTH). We omit the print functions for brevity; they can be found in [10].

Our element ErrorStatsReq sends a periodic ERROR_STATS_REQ request, triggered by a timer. The function that prepares the message is the following:

```
1  void
2  ErrorStatsReq::sendErrorStatsReq() {
```

```
3      const unsigned char dst[6] = {0x00, 0xB0, 0x52, 0x00, 0
       x00, 0x01};
4
5      static_assert(Packet::default_headroom >= sizeof(click_
       ether));
6      WritablePacket *q = Packet::make(Packet::default_
       headroom, NULL, sizeof(click_ether) + sizeof(click_hp
       _av_header) + sizeof(click_hp_av_error_stats_req), 0)
       ;
7      if (!q) {
8          click_chatter("[ErrorStatsReq] cannot make packet
           !");
9          return;
10     }
11
12     click_ether *e = (click_ether *) q->data();
13     q->set_ether_header(e);
14     memset(e->ether_shost, 0x00, 6);
15     memcpy(e->ether_dhost, dst, 6);
16     e->ether_type = htons(ETHERTYPE_HP_AV);
17
18     click_hp_av_header *hpav = (click_hp_av_header *) (e +
       1);
19     hpav->MMType = htons(ERROR_STATS_REQ);
20     hpav->version = 0;
21
22     click_hp_av_error_stats_req *error_req = (click_hp_av_
       error_stats_req *) (hpav + 1);
23     memcpy(error_req->macaddr, &_dst, 6);
24     error_req->link_id = _prio;
25     error_req->direction = _dir;
26     error_req->control = 0;
27     const unsigned char oui[3] = {0x00, 0xB0, 0x52};
28     memcpy(error_req->oui, oui, 3);
29
30     q->timestamp_anno().assign_now();
31     output(1).push(q);
32  }
```

We create a new packet q with the existing function `WritablePacket *q = Packet::make()` (line 6). The function takes as input the total size of the packet, as measured by the headers we need to use. Then, we fill the packet with the appropriate Ethernet and HP AV headers (lines 12–30). We push the packet to output 1 of our element (line 31). This output is connected to a **Queue**, which itself is connected to the **ToDevice** element (see Figure 4.6). Our request is now forwarded to the Ethernet interface.

4.8.3   Compiling and Running New PLC Click! Elements

When the code for Click! elements is ready, we need to recompile Click! for the operating system we are using for testing. We have to add our new elements in the Click! path of our source code under the folder elements[11]. We have to make sure

---

[11] See the folder at github.com/kohler/click/tree/master/elements.

that our elements will be included in the compilation. We create a new directory under elements, elements/plc. To add our PLC elements in the compilation list, we add the following line in the file configure.in:[12] ELEMENTS_ARG_ENABLE(plc, [include plc elements], yes). Then, we run the command make elemlist in the Click! root path for the new elements to be added. To finally compile Click!, we run the ./configure and make install commands.

To run the Click! script plc_elem.click, we should first execute the command click-align plc_elem.click -o plc_elem_final.click, which adds required **Align** elements for machines that don't support unaligned accesses into an updated script that we call texttplc_elem_final.click. Each **Align(MODULUS, OFFSET)** input packet is aligned so that its first byte is OFFSET bytes off from a MODULUS-byte boundary. MODULUS must be 2, 4, or 8. After alignment, we can run the script as click plc_elem_final.click.

4.9 Extension of the Tools to Other Vendors

In this chapter, we have focused on existing tools for Qualcomm-based PLC devices. We have also showed how to develop new tools, interacting with PLC devices using MMs. We now extend this work and consider different vendors and devices supporting the newest HomePlug AV2 specification. The major vendors of HomePlug devices are Qualcomm and Broadcom. Hence we now turn our attention to HomePlug AV2 developed by Broadcom. We have used the D-Link Gigabit Starter Kit DHP-701AV,[13] and have explored how to develop tools for or interact with such devices.

A first step when testing a new PLC device is to look for the vendor's software releases. Typically, all vendors supply software for troubleshooting and configuring devices. The most important functionality of such software is modifying the network membership key to ensure security and privacy in the network. By using the vendor's software and capturing the packet flow to the Ethernet interface of the PLC device, we can infer the vendor-specific MMs and reuse them in new tools for educational purposes, research, or hybrid network systems.

We now describe how to infer vendor-specific MMs in detail, with the example of the DHP-701AV device. We install the software from D-Link on a Windows machine[14]. In addition, we have to install Wireshark [9],[15] an open source program that captures packet flow from any interface. We then connect one of the PLC devices to the Windows machine and a second PLC device to another computer. By default, the devices form a network as they share the same default membership key. After installing software and

---

[12] See github.com/kohler/click/blob/master/configure.in.

[13] DHP-701AV uses Broadcom's chipset BCM60500; see www.smallnetbuilder.com/lanwan/lanwan-reviews/32716-d-link-dhp701av-powerline-av2-2000-gigabit-network-extender-kit-reviewed.

[14] D-Link released software available for Windows OS: ftp://ftp2.dlink.com/SOFTWARE/PLC_Utility/DLINK_PLC_UTILITY_v5.02B02.zip.

[15] In Linux, we can use Wireshark, tcpdump, or tshark.

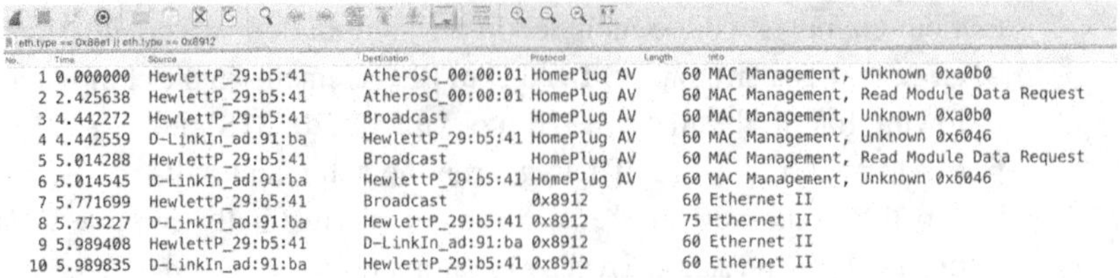

eth.type == 0x88e1 || eth.type == 0x8912

No.	Time	Source	Destination	Protocol	Length	Info
1	0.000000	HewlettP_29:b5:41	AtherosC_00:00:01	HomePlug AV	60	MAC Management, Unknown 0xa0b0
2	2.425638	HewlettP_29:b5:41	AtherosC_00:00:01	HomePlug AV	60	MAC Management, Read Module Data Request
3	4.442272	HewlettP_29:b5:41	Broadcast	HomePlug AV	60	MAC Management, Unknown 0xa0b0
4	4.442559	D-LinkIn_ad:91:ba	HewlettP_29:b5:41	HomePlug AV	60	MAC Management, Unknown 0x6046
5	5.014288	HewlettP_29:b5:41	Broadcast	HomePlug AV	60	MAC Management, Read Module Data Request
6	5.014545	D-LinkIn_ad:91:ba	HewlettP_29:b5:41	HomePlug AV	60	MAC Management, Unknown 0x6046
7	5.771699	HewlettP_29:b5:41	Broadcast	0x8912	60	Ethernet II
8	5.773227	D-LinkIn_ad:91:ba	HewlettP_29:b5:41	0x8912	75	Ethernet II
9	5.989408	HewlettP_29:b5:41	D-LinkIn_ad:91:ba	0x8912	60	Ethernet II
10	5.989835	D-LinkIn_ad:91:ba	HewlettP_29:b5:41	0x8912	60	Ethernet II

Figure 4.7 Wireshark trace: understanding the MM structure of different chipset vendors, such as Broadcom.

devices, we open the D-Link software and Wireshark. We start a Wireshark packet capture on the interface to which the PLC device is connected. Then, we start interacting with D-Link software. For example, we can change the membership key of the device, we can upgrade the firmware of the device, we can scan for all devices in the network, and so on. A screenshot of these capabilities and a packet capture example are shown in Figures 4.4 and 4.7, respectively. Our application discovers the two connected devices, which we have named "PLC-A," and "PLC-B." Our local device has the MAC address c4:12:f5:ad:91:ba, and the remote one has c4:12:f5:ad:91:b9. In our trace, we see messages from both devices when we request firmware information and only from the local device when we request network and link capacity discovery.

In Wireshark, we set a filter to display only packets of EtherType HomePlug AV (0x88e1) or a new EtherType 0x8912, which we discovered that these devices are using, as we explain later. To present our findings, we refer to the packet numbers at the first column of the capture in Figure 4.7. In Packets 1 and 2 captured, we observe that software first tries to send MMs to an Atheros MAC address, because D-Link has products with both Qualcomm/Atheros and Broadcom chipsets and a network might consist of heterogeneous devices. The PLC devices do not respond to these MMs (due to Broadcom chipset), hence the software sends new MMs with HomePlug AV EtherType to broadcast MAC address (Packets 3 and 5). We observe that the D-Link device replies (Packets 4 and 6). D-Link software has discovered the local PLC device directly connected to this computer. Then, the software broadcasts Packets 7 and 9, which are two different MMs with EtherType 0x8912. This is a vendor-specific EtherType that we discovered by repeating operations in the D-Link software. Nevertheless, we can infer the nature of these new messages. Packet 7 has MMType 0xa070, and the reply to this MM, Packet 8, contains the payload "2DHP700AVA1_FW101b01_FWDate_12_Mar_2015_CS_31810417" (see Figure 4.8), which is the firmware version. Hence the D-Link device uses MMType 0xa070 for retrieving firmware version.

We discover that the PLC devices support a very important functionality for heterogeneous networks, that is, capacity estimation. Packet 9 has MMType 0xa02c, and its reply Packet 10, whose byte structure is shown in Figure 4.9, contains the number of neighboring PLC devices (one in our case), their MAC address (c4:12:f5:ad:91:b9), and the capacity of both traffic directions (bytes 0xa265 and 0x4565). Packet 10 first starts with the MAC header, with source address the Ethernet of our PC, destination the MAC

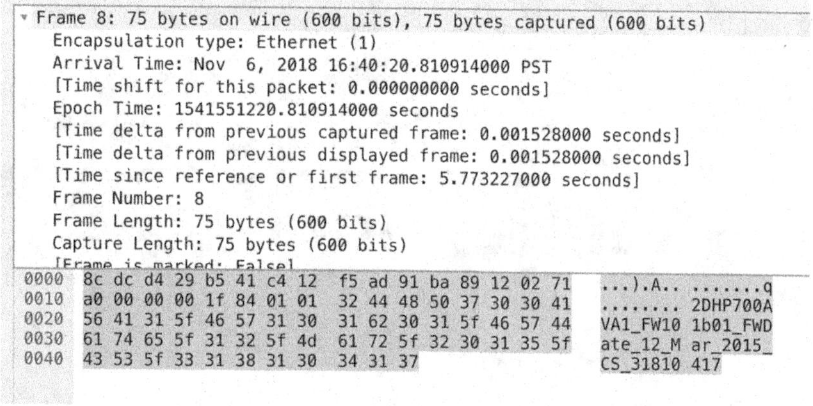

```
▼ Frame 8: 75 bytes on wire (600 bits), 75 bytes captured (600 bits)
    Encapsulation type: Ethernet (1)
    Arrival Time: Nov  6, 2018 16:40:20.810914000 PST
    [Time shift for this packet: 0.000000000 seconds]
    Epoch Time: 1541551220.810914000 seconds
    [Time delta from previous captured frame: 0.001528000 seconds]
    [Time delta from previous displayed frame: 0.001528000 seconds]
    [Time since reference or first frame: 5.773227000 seconds]
    Frame Number: 8
    Frame Length: 75 bytes (600 bits)
    Capture Length: 75 bytes (600 bits)
    [Frame is marked: False]
0000  8c dc d4 29 b5 41 c4 12  f5 ad 91 ba 89 12 02 71   ...).A.. ......q
0010  a0 00 00 00 1f 84 01 01  32 44 48 50 37 30 30 41   ........ 2DHP700A
0020  56 41 31 5f 46 57 31 30  31 62 30 31 5f 46 57 44   VA1_FW10 1b01_FWD
0030  61 74 65 5f 31 32 5f 4d  61 72 5f 32 30 31 35 5f   ate_12_M ar_2015_
0040  43 53 5f 33 31 38 31 30  34 31 37                  CS_31810 417
```

Figure 4.8 Wireshark trace: firmware version.

of the local PLC-B device, and Ethertype 0x8912. Then, we have the byte 0x02, which hints that the MM version is for HomePlug AV2, and the MMType (in reverse network byte order 0xa02d). After the MMType, we have seen the bytes 0x0000001f84 being repeated for every proprietary MM. It is a vendor-specific header for the MM data. These bytes are followed by 0x01 or 0x02, if the MM is a request or reply, respectively. Packet 10 is a reply, hence we have the sequence of bytes 0x0000001f8402. Then we observe that the sequence is followed by the byte 0x01, which is the number of PLC neighbors. (If we do not connect the second device, this number is 0x00.) The software shows the capacity per PLC neighbor in Mbps. In our example, capacity is measured as 1442 Mbps, as we see in Figure 4.9. We observe that the hex representation 0xa265 could correspond to number 0x5a2 in hex, which is 1442 in decimal, and the hex 6 could signify the Mbps metric (10^6 bits). We verified this assumption by testing with different link capacities larger than 1000 Mbps and smaller than 100 Mbps. Hence the D-Link device uses MMType 0xa02c for network discovery and link quality estimation. To run new measurements and retrieve real time capacity, we can click the refresh button on the software or develop our own tools sending periodic capacity updates.

As we discuss in Section 4.5.2, security and privacy have to be guaranteed in PLC. We have shown how to set security keys with Atheros/Qualcomm devices. Now, we move to the Broadcom vendor and explore the MMs to set the NMK. In Figure 4.10, we show that with this vendor, we can set the 16-byte NMK by MMs with MMType 0xa018. In this example, we set the NMK to ef 02 85 2d 83 13 65 93 48 86 d6 b5 67 27 a9 42. The device confirms the new NMK by D-Link software and an MM of MMType 0xa019.

Table 4.6 summarizes the MMs we have discovered through this software and their corresponding MMType. We observe that by using the vendor's software, which is provided for most PLC devices, we can learn new vendor-specific MMs and use them for network configuration and measurements.

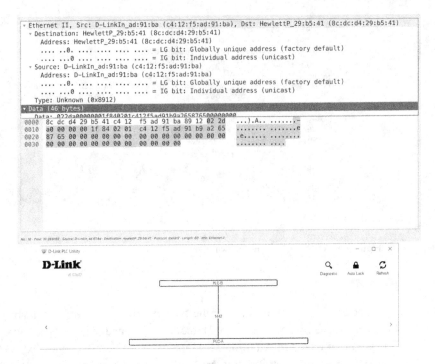

Figure 4.9 Wireshark trace: neighbor discovery and capacity estimation. D-Link software: capacity interpreted from message captured by Wireshark.

Table 4.6 D-Link/Broadcom vendor-specific MMs, as discovered from vendor's software

MMType (request/reply)	Functionality
0xa018/0xa019	Change device's security key
0xa070/0xa071	Get firmware version
0xa02c/0xa02d	Get connected devices and link capacity
0xa054/0xa055	Reset to factory defaults
0xa024/0x6049	Get PIB info with offset

Note. The reply to a request MM always sets the least significant bit of MMType to 1.

4.10   Summary

We have introduced our experimental framework for commercial PLC devices. We have explained in detail how to configure the network for security and reliable performance. In addition, we have provided guidelines for retrieving various statistics on PHY- and MAC-layer performance. The command line tools introduced here are open source and can be used by most Linux and Mac operating system users. Various functionalities of the aforementioned tools are also available with other operating systems, usually through software released by the vendor. The user can also deploy new tools for any operating system by generating PLC MMs with the appropriate software. The operation of our guidelines is based on simple management messages exchanged between

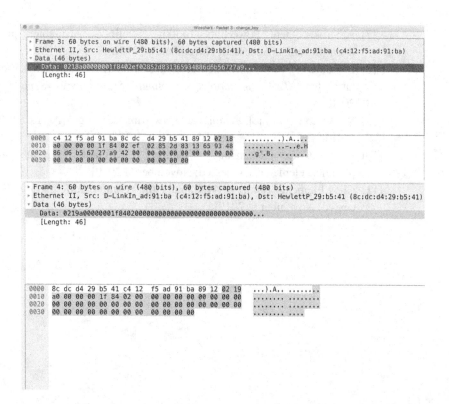

Figure 4.10 Wireshark trace: modification of NMK for Broadcom chipset and confirmation from the device.

the Ethernet and PLC interfaces. In the next two chapters, we show how to use this experimental framework in practice for measuring and configuring PLC performance.

References

[1] PC Engines. [Online]. Available: pcengines.ch/alix.htm. [Accessed: November 6, 2019].

[2] OpenWrt. [Online]. Available: openwrt.org/. [Accessed: November 6, 2019].

[3] INT6300. [Online]. Available: www.electronnics.com/data/INT6300.pdf. [Accessed: November 6, 2019].

[4] AR7400. [Online]. Available: www.codico.com/fxdata/codico/prod/media/ Datenblaetter/AKT/AR7400+AR1500%20Product%20Brief.pdf. [Accessed: November 6, 2019].

[5] *IEEE Standard for Broadband over Power Line Networks: Medium Access Control and Physical Layer Specifications*, IEEE Standard 1901-2010.

[6] X. Carcelle and F. Fainelli, "FAIFA: A first open source PLC tool," 2008, [Online]. Available: fahrplan.events.ccc.de/congress/2008/Fahrplan/events/2901. en.html. [Accessed: November 6, 2019].

[7] Atheros Open Powerline Toolkit. [Online]. Available: github.com/qca/open-plc-utils. [Accessed: November 6, 2019].

[8] E. Kohler, R. Morris, B. Chen, J. Jannotti, and M. F. Kaashoek, "The Click modular router," *ACM Transactions on Computer Systems*, vol. 18, no. 3, pp. 263–297, 2000.

[9] Wireshark. [Online]. Available: www.wireshark.org/. [Accessed: November 6, 2019].

[10] PLC Click Elements. [Online]. Available: github.com/christinavl/plc-click-elements. [Accessed: November 6, 2019].

[11] Click Elements. [Online]. Available: github.com/kohler/click/wiki/Elements. [Accessed: November 6, 2019].

Part II

How Does PLC Perform?

5 PHY-Layer Performance Evaluation

5.1 Introduction

In this chapter, we evaluate the spatio-temporal variations of end-to-end PLC performance of single links, and we propose practical link-metric estimation methods that can be used in hybrid networks (such as IEEE 1905). We investigate PHY and MAC layers of PLC with a testbed of more than 140 links and devices from two different vendors and two different HomePlug specifications.

In Section 5.2, we discuss PLC connectivity and topology and investigate the spatial variation of PLC. Measurements show that PLC links can be highly asymmetric and this has two consequences: (1) link metrics should be carefully estimated in both directions and (2) predicting which PLC links will be good is challenging. The spatial-variation study yields important observations for large-scale PLC deployments and connectivity.

We explore the temporal variation of the PLC channel in Section 5.3, and distinguish three different timescales for capacity variability, called "invariance," "cycle," and "random". Each timescale yields fundamental insights for exploiting PLC to its fullest extent, for predicting end user performance variability, and for efficiently updating link metrics (e.g., high-frequency probing yields accurate estimations but high overhead). We discuss that good links might vary much less often than bad ones and that capacity-estimation overhead can be reduced by probing good links less often. Section 5.4 compares Wi-Fi and PLC link variability.

In Section 5.5, we discuss a capacity-estimation technique for PLC links and evaluate it under various settings; the proposed capacity estimator is the bit-loading estimate BLE (introduced in Section 2.7), which is included in every PLC frame header. We examine the ideal packet size and priority class of probing traffic for capacity estimation. We also explore the accuracy of the capacity-estimation technique by designing a load-balancing algorithm for hybrid Wi-Fi/PLC single-hop networks, using the Click! modular router introduced in Section 4.8.

Capacity estimation might not be sufficient as a link metric in hybrid networks, as quality of experience also depends on retransmissions and error rates. To include metrics related to packet errors and delay, in Section 5.6, we examine the PLC retransmission procedure and how link metrics are affected by contention. There we investigate both broadcast and unicast methods for evaluating link quality, and we present a technique to address potential link-metric sensitivity to background traffic. We also measure the PB error rate (PB_{err}) that is defined in the IEEE 1901 standard. Both metrics BLE and

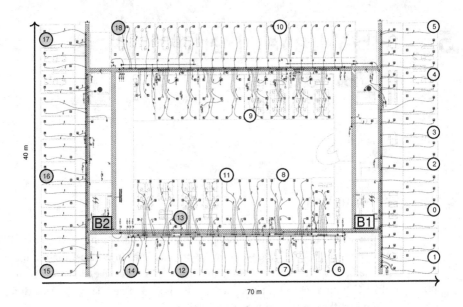

Figure 5.1 The electrical plan and the stations (0–18) of our testbed. Stations marked with the same color (white or gray) belong to the same network and are connected to the same distribution board (either B1 or B2).

PB_{err} are compliant with the IEEE 1905.1 standard for hybrid networks. At the end of every section, we give instructions on how to reproduce the measurements using the experimental framework built in Chapter 4. We summarize our guidelines for link-metrics estimation in Section 5.7.

5.2   Spatial Variation of PLC

The spatial variation of PLC is interesting for future deployment and popularity of PLC and implementations of hybrid networks. We first discuss PLC topology and connectivity, then the predictability of link quality with respect to electrical cable length.

5.2.1   Topology and Connectivity

Our experiments are conducted on a testbed spread over a floor of a university building at École polytechnique fédérale de Lausanne (EPFL), Switzerland. The map of our testbed and the electrical plan of the floor are shown in Figure 5.1. We observe that there are two electrical distribution boards, denoted by B1 and B2. Stations 0–11 are connected to B1, and 12–18 are connected to B2. We use two types of hardware with HomePlug AV and AV500 specifications. Compared to AV described in Section 2.4, AV500 extends the bandwidth to 1.8–68 MHz and maintains the rest of the AV PHY-layer mechanisms unchanged. The details on configuring PLC devices and measuring statistics can be found in Chapter 4. In the following, we present our findings, with respect to the topology and connectivity of PLC.

As mentioned in Section 3.7, a PLC network requires a central authority to operate, called the CCo. Our floor has two distribution boards that are connected with each other at the basement of the building. This means that the cable distance between the two boards (more than 200 m) makes the PLC communication between two stations at different boards challenging. Owing to the two distribution lines, none of the stations can communicate with all stations and become the CCo, yielding at least two PLC networks. Hence, we create two PLC networks, represented with different colors in Figure 5.1, with CCo's at Stations 1 and 15. We set the CCo statically, as described in Section 4.5. Our finding suggests that in large-scale deployments, a CCo per distribution board (or proxy CCo) must be provisioned due to the potential attenuation introduced between different distribution boards.

⚄ **Experimental Guideline**   To create the two PLC networks, we have first to set different keys for each network. For every node, we can run the following commands:

```
1    int6k -p <PIB_FILE> <MAC_ADDRESS>  # Retrieve the PIB
         file from the PLC interface
2    modpib -N <NMK> -v <PIB_FILE> # Modify NMK
3    int6k -P <PIB_FILE> <MAC_ADDRESS> # Update the PIB file
         of the PLC interface
```

The NMK value in line 2 above has to be 16 bytes and different for the two sets of nodes 1–11 and 12–18 in our example of Figure 5.1. After setting the NMK, we need to configure the CCos of the networks. In our testbed, we have determined the CCo by observing how different networks are formed after restarting all devices. The devices close to the border between the networks for example Station 12, could connect to both networks. However, for a stable performance, we allocate devices to each network based on the electrical distribution board to which it belongs. We can execute the following commands:

```
1    plctool -i <ETH_INTERFACE> local -p <PIB_FILE>  #
         Retrieve the PIB file from the PLC interface
2    modpib -C <MODE> -v <PIB_FILE> # Modify CCo mode of
         <PIB_FILE>, <MODE>: 0 Auto, 1 Never, 2 Always
         CCo
3    plctool local -i <ETH_INTERFACE> -P <PIB_FILE> #
         Update the PIB file of the PLC interface
```

In line 2 above, we set the MODE to 2 for the two CCo Stations 1 and 15, and we set MODE to 1 for the rest of stations. Finally, we reboot all the devices. ⚄

In terms of connectivity and topology, we observe that each of the two PLC networks forms an almost fully connected graph. This yields very good connectivity for PLC networks that belong to a single distribution board. In addition, we observe that there is some connectivity between stations at different boards, for example Station 11 with 13 and 12. We did not manage to verify whether this connection is due to a direct path between cables (which meet in the basement of our building, where two boards are connected) or to electromagnetic interference from neighboring electrical cables on the testbed floor. For this reason, in the following section, we investigate link quality as a function of cable distance, excluding these few cases of links between different distribution boards.

5.2.2   Asymmetry in PLC Links and Consequences

We explore the spatial variation of PLC, as it is important for predicting coverage and the good locations for PLC stations, and for implementing link metrics. We find that PLC can be highly asymmetric, and this should be considered when estimating link metrics or designing new networks.

A very important characteristic of power line channels is that they exhibit performance asymmetry, that is, capacity can differ significantly between the two directions of the link. In the experiments that we run (both with AV and AV500), we observe a performance asymmetry of more than 1.5x in approximately 30 percent of stations pairs in our testbed. Figure 5.2 presents typical examples of these links, for which the throughput in one direction is less than 60 percent of the throughput in the opposite direction. By re-conducting the experiments with AV500 devices, we verify that the asymmetry is not due to the hardware. Link asymmetry in PLC has also been observed by Murty et al. [1]. We attribute this asymmetry to a high electrical load (e.g. one or more appliances with much higher impedance than the cable's impedance) that exists near one of the two stations. In this case, the channel cannot be considered as symmetric, and the two transmission directions experience different attenuations.

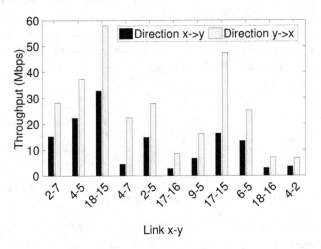

Figure 5.2 Available-throughput asymmetry in PLC links.

We next present a spatial variation study, where we use both AV and AV500. Figure 5.3 provides the available UDP throughput of single links as a function of the cable distance between the source and the destination of the traffic from a single experiment. There is a clear degradation of throughput as the distance increases. However, because of the diversity in positions and types of connected appliances, there is a large range of possible throughputs at any specific distance. We observe that small distances (<30 m)

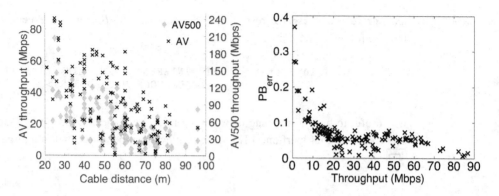

Figure 5.3 Available throughput versus cable distance between source and destination for all links of the testbed (left). PB_{err} versus available throughput for AV (right).

guarantee good links, but that large distances (30–100 m) can yield either good or bad links. By comparing AV and AV500, we observe that AV500 enables some links with no AV connectivity to still have a nonzero throughput, but with severe asymmetries (e.g., link 10-2 has a 10x capacity difference between two directions).

To further explore the causes of PLC attenuation and variability, we run experiments with two stations connected by a long electrical cable and without any devices attached. We notice that the attenuation in an up to 70 m cable causes a throughput drop of at most 2 Mbps. Therefore the attenuation in the realistic topology is caused by the multipath nature of the channel and not by cable length, as explained in Section 2.2. By plugging electrical appliances in this isolated test, we observe that asymmetry was introduced, as found also by Murty et al. [1].

From the above, spatial variation of PLC can depend on diverse factors: (1) the structure of the electrical network, that is, the circuit breakers and distribution boards; (2) the electrical appliances attached and their position on the grid; and (3) (weakly) the cable-distance between stations. The channel can be very asymmetric and this is a key feature for studying spatial variability.

To optimize performance not only in terms of throughput but also in terms of delay, hybrid networks need some estimation of the retransmissions a frame suffers due to channel errors or collisions. We evaluate the relationship of the metric PB_{err} with the available throughput. Figure 5.3 illustrates PB_{err} versus the available throughput for all the links of our testbed. It shows that PB_{err} decreases as throughput increases, as expected. However, because the tone maps are updated based on this metric, some average links might have PB_{err} lower than the best links of the testbed. We further study the PB_{err} metric in Section 5.6 and show that PB_{err} can be used to predict the expected number of retransmissions due to errors.

⌨ Experimental Guideline   To repeat this experiment and measure application throughput, you can run the Linux utility `iperf` between two PLC devices. To test for asymmetry, you should run two back-to-back experiments between two PLC devices as an `iperf` client and server, respectively. To measure the PB_{err}, we can execute for every link the following command:

```
1           int6kstat -i <ETH_INTERFACE> -d <LINK_DIRECTION> -s
              <LINK_ID> -p <PEER_NODE_MAC_ADDRESS>
```

It is useful to have a mapping between the IP address and the PLC MAC address for every node so that iperf and PHY measurements can be automated in a large testbed.

5.3   Temporal Variation of PLC

We discuss the timescales within which the channel varies. These timescales were introduced for channel modeling and simulation by Sancha et al. [2] (from which we borrow the terminology to name these timescales). Here, we study them experimentally. This study differs from the work by Sancha et al. [2] in that we examine the channel quality from an end user and practical perspective, exploring metrics that affect the end-to-end performance.

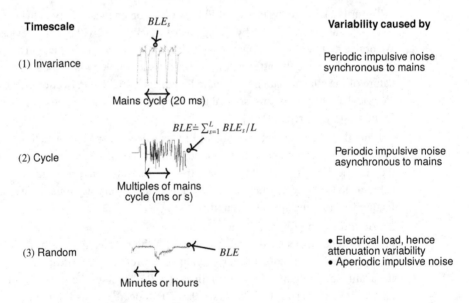

Timescale	BLE_s	Variability caused by
(1) Invariance	Mains cycle (20 ms)	Periodic impulsive noise synchronous to mains
(2) Cycle	$BLE \doteq \sum_{s=1}^{L} BLE_s / L$ Multiples of mains cycle (ms or s)	Periodic impulsive noise asynchronous to mains
(3) Random	BLE Minutes or hours	• Electrical load, hence attenuation variability • Aperiodic impulsive noise

Figure 5.4 BLE (capacity) temporal variation. BLE_s denotes BLE at tone map slot s.

We examine separately the two main components of channel modeling, i.e., the variation of noise generated by the attached electrical appliances to the grid and the variation of the channel transfer function (or attenuation). We use BLE to investigate the main properties of PLC channels, by using existing commercial devices. We show that BLE can reflect the channel quality and the fundamental features of PLC channel modeling explained in Section 2.2. We first focus on noise generated by electrical appliances. It has been shown by measurements that the noise level varies across subintervals of the mains cycle, which yields the first scale governing PLC temporal variation (scale (1)) [3]. Owing to the periodic nature of the mains, this noise also varies in a scale of multiples of the mains cycle, which results in another timescale for the temporal variation (scale (2)).

We next focus on attenuation. As discussed in Section 2.2, attenuation is introduced mainly due to impedance mismatches in the transmission line (electrical cable), which are created by connected appliances. As expected, this attenuation changes when the structure of the electrical network changes, hence on a scale of minutes or hours (scale (3)). This variability strongly depends on appliances usage and on switching the appliances, as this creates aperiodic impulsive noise and different attenuation in the channel. Figure 5.4 illustrates the three timescales and the factors that cause variability. The three timescales are as follows.

(1) **Invariance scale:** subintervals of the mains cycle, such as the tone map slots;
(2) **Cycle scale:** multiples of the mains cycles – depends on the noise produced by appliances;
(3) **Random scale:** minutes or hours – related to connection or switching of electrical appliances and depends on human activity.

We next examine each timescale of PLC variability. We introduce our variables, starting with some notation. For the invariance scale, we use the term tone map slots for the subintervals of the mains cycle, as we can measure the channel quality with respect to tone map slots by using PLC devices. Let L be the total number of tone map slots of the mains cycle (in practice with IEEE 1901 standard [4], $L = 6$), with each slot s having a duration T_s, so that the total slot duration $\sum_{s=1}^{L} T_s$ is equal to half of a mains period, which is 20 ms for 50 Hz current in Europe. Let BLE_s, $1 \le s \le L$, denote the BLE of tone map slot s. To study the channel with respect to the three scales defined above, we assume that time t is discrete, with one time unit having real-time duration equal to the mains cycle: $BLE_s(t)$ is the BLE for the sth tone map slot of the tth mains cycle. The process $BLE_s(t)$ is different for each link, and its distribution can be time varying over the random scale for a specific link, owing to the different types of operating appliances and to different channel transfer functions.

5.3.1 Invariance Scale

The invariance scale of BLE is affected by the noise levels that appliances produce at different subintervals of the mains cycle, and it has direct consequences on estimating link metrics. Most tests show that noise has varying levels over different tone map slots, thus confirming in practice the existence of periodic impulsive noise, synchronous to the mains (see Section 2.2). Figure 5.5 shows the instantaneous BLE_s from captured frames in typical examples of good and average links. We observe that in HomePlug AV, the total duration of the six tone map slots is equal to half of the mains cycle, thus BLE_s change periodically, with a period of 10 ms. Each PLC MPDU uses a different BLE_s, depending on in which tone map slot s its transmission takes place.

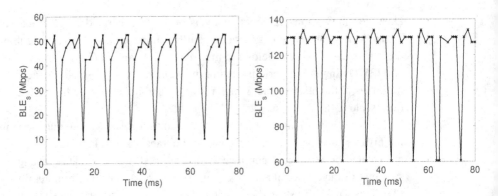

Figure 5.5 Invariance-scale variation of BLE from captured PLC frames of two representative links, 6-1 left and 0-2 right. The frames and BLE metric are captured using the sniffer mode of PLC devices, as explained in Section 4.7.3.

The invariance timescale is crucial for capacity estimation in PLC. With the examples of Figure 5.5, we observe that there might be significant variations within the mains cycle, even for good and average links. Thus, link metrics should be estimated or averaged over all $L = 6$ tone map slots of the periodic modulation scheme of PLC.

5.3.2   Cycle Scale

We examine the average time during which the quality of the links is preserved in the cycle scale. This sheds light on the average length of probing intervals for link metrics, as there exists a trade-off in probing: too large intervals might yield an inaccurate estimation, whereas too small intervals can generate high overhead.

We conduct experiments that last 4 min, over all links of the testbed. We send saturated throughput and measure BLE using MMs (see Section 4.6). During each experiment, we request BLE_s, $1 \leq s \leq L$, every 50 ms, as this is the fastest rate at which we can currently send MMs to the PLC chip. As we need to avoid random changes in the channel due to switching electrical appliances, all the experiments of this subsection are conducted during nights or weekends. For the cycle-scale variations of the channel, we assume that the electrical network structure is fixed.

Here we evaluate the cycle-scale variation by using $BLE \doteq \sum_{s=1}^{6} BLE_s/6$, in other words, the average BLE over all tone map slots. We compare the performance between good and bad links.[1] Figure 5.6 presents the variation for typical good and bad links of our testbed. For HPAV, the experiments have been repeated two times for each link, 2–4 months apart. Observe that depending on their quality, links can exhibit different behaviors:

Bad links: Bad links, for example, 11-4 and 6-5, tend to modify the tone maps much

---

[1] The classification of the links based on their capacity depends on the PLC technology; consequently, we do not use strict thresholds for this characterization.

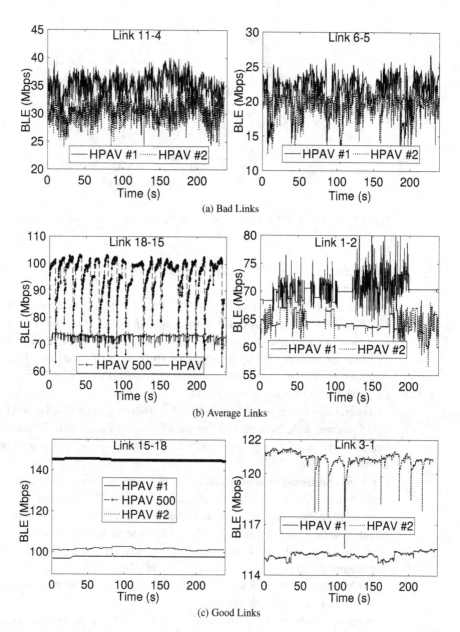

Figure 5.6 Examples of BLE cycle-scale variation for links of diverse qualities. Experiments for HPAV have been repeated two times, 2–4 months apart.

more often than good links do. Moreover, they yield a standard deviation of BLE significantly higher than good links.

Average links: Average links, for example, 18-15 and 1-2, vary less often than bad links, and might preserve their tone maps for a few seconds. During periods when av-

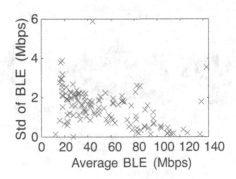

Figure 5.7 Cycle-scale variation of BLE with respect to the link quality.

erage links vary often, the standard deviation of BLE can be high (e.g., 1-2), depending on the channel conditions.

Good links: The tone maps of good links can be valid for several seconds, e.g., link 15-18. Good links, such as link 3-1, that often update the tone maps have insignificant increments or decrements, for example, of up to 1 percent, or have impulsive drops of BLE, for example, of up to 5 percent (between 116 Mbps and 122 Mbps in the rightmost figure), with the channel-estimation algorithm needing a few time steps to converge back to the average BLE value.

Asymmetry in temporal variability: By observing links 15-18 (good link, left) and 18-15 (average link, left), we find that the asymmetry discussed in Section 5.2 can translate not only in an average-performance asymmetry, but also in a temporal-variation asymmetry.

Channel-estimation algorithms: The temporal variation of link 15-18 is similar with AV and with AV500. By noticing the impulsive BLE drops in link 18-15 and by comparing AV with AV500, we detect a feature of the channel-estimation algorithm that might be vendor specific: the AV500 performance oscillation shows that, when bursty errors occur, the estimation algorithm returns very low values of BLE. This uncovers that temporal variation in PLC link quality can also depend on the channel-estimation algorithm, and future work should focus on comparing link-metric estimations for different vendors and technologies.

We next corroborate the above findings over all links of our testbed. Figure 5.7 shows the standard deviation of BLE over our experiment (250 s) for all links. We observe that good links have a lower BLE variability compared to bad links. Although some good links might update BLE at a similar frequency as bad links (~ 100 ms), as we discussed above, these links tend to have small increments and decrements of BLE, thus yielding a stable average performance over minutes and a low BLE standard deviation.

To reduce overhead, in cycle scales, that is, seconds or minutes, good links should be probed less often than bad links. The cycle-scale variation unveils how link metrics should be updated depending on their quality. It also shows that PLC links are quite stable, providing at most 6 Mbps standard deviation over short timescales.

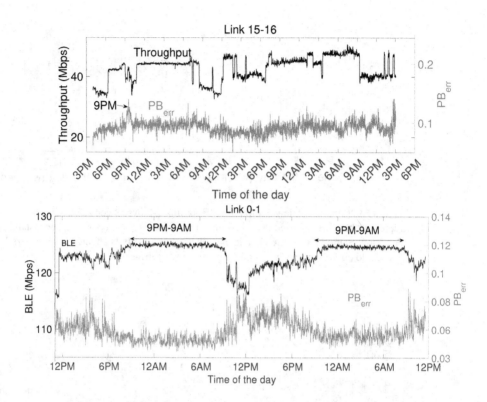

Figure 5.8 Random-scale variation of PLC over a total duration of 2 days. Metrics are averaged over 1-min intervals. Every night at 9:00 pm, all lights are turned off in our building, leading to a channel change for PLC.

5.3.3   Random Scale

In Section 5.3.2, we observe that during timescales of seconds, PLC capacity does not typically vary much, with a standard deviation of throughput up to 6 Mbps. We now look at longer timescales, that is, in terms of minutes and hours, with two goals: (1) to examine whether some links could be probed at a slow rate, thus reducing overhead, and (2) to characterize the variability of PLC performance in the presence of high and low electrical loads. To study the channel quality variation over the random scale, we run tests over long periods, that is, two days and two weeks, for various links. During these tests, we measure throughput, BLE, and PB_{err} every second. Figures 5.8–5.10 show the results of our measurements. Our observations are as follows.

Link quality versus time: The variation of BLE is governed by the electrical load. The larger the number of switched-on devices is, (e.g., at working hours) the larger the attenuation is and the lower BLE is, as discussed in Section 2.2.

Link quality versus variability: In Figures 5.9 and 5.10, which represent a good and a bad link, observe the differences in the y-axis scales. For a given link, the random-scale variation of BLE strongly depends on the noise of the electrical devices attached, and it is higher when BLE is lower. High standard deviation of BLE and low average values

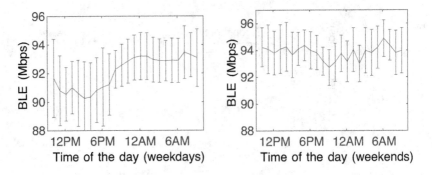

Figure 5.9 Random-scale variation of BLE for link 1-8 over 2 consecutive weeks. Lines represent the BLE averaged over the same hour of the day, and error bars show standard deviation. Station 8 is located in a lounge, hence link quality is affected by the operation of household appliances, such as a microwave oven or refrigerator.

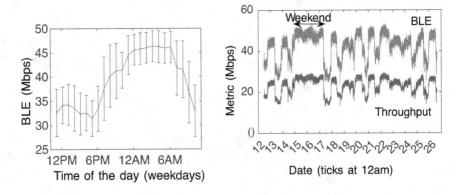

Figure 5.10 Random-scale variation of BLE for link 2-11 over 2 consecutive weeks. The link quality is stable during the first weekend, but not during the second, due to an event held in our building and due to high electrical activity.

imply that more devices are switched on, therefore more noise is produced. They also imply that devices are switched on/off more often, creating impulsive noise phenomena. The standard deviation of BLEs is very small for good links; it increases as the link quality decreases.

Link probing: Good links seem to exhibit a negligible standard deviation, which implies that they can be probed every minute or hour, depending on the time of the day.

To conclude, PLC links seem to exhibit low temporal variability in short time-scales. Good and average links are quite stable over time, hence they can reliably serve bandwidth-hungry applications. The quality of bad links has high variance in the random scale. The random-scale variation of PLC can be predicted by learning the electrical-load patterns in the building.

📖 **Experimental Guideline** To repeat the experiments in this section, we suggest two methodologies. To study capacity over the invariance and cycle scales, we need per-MPDU measurement granularity. Hence, for Sections 5.3.1 and 5.3.2, run the Linux utility `iperf` between two PLC devices for saturated traffic, and then enable the sniffer mode of PLC devices to capture all the SoF delimiters for MPDUs. We can use the following command and save the output to a file:

```
1    faifa -m -i <ETH_INTERFACE>
2    Choose the frame type (Ctrl-C to exit): 0xa034
3    Frame: Sniffer Mode Request (0xA034)
4    Sniffer mode?
5    0: disable
6    1: enable
7    2: no change
```

Select mode 1 above to enable sniffer mode. At the end of the experiment, you can re-run the command above to disable the sniffer mode with option 0, to avoid frame capturing overheads on the device. It is best to run this command at the station that transmits or receives the data, as other stations might not be able receive all the MPDUs due to channel quality differences. Then, we have to compute timestamps and BLEs from frames with delimiter type 1, which represents data frames:

```
1    Direction: Rx
2    System time: 26657900886
3    Delimiter type: 1
4    STEI: 03
5    DTEI: 06
6    Link ID: 01
7    HPAV version 1.0: No
8    HPAV version 1.1: No
9    Bit loading estimate: 26
10   HPAV frame length: 530
11   MPDU count: 1
```

Recall that BLE encoded as $BLE = (32 + mant) \times 2^{exp-4} + 2^{exp-5}$, where mantissa (*mant*) is the first 5 bits and exponent (*exp*) is the last 3 bits of the BLE value shown above in line 9. We create a time series of timestamps and BLE. Then, we clean the data as the capturing interface might have skipped a few frames. For every sequence of six BLE values (remember that BLEs are periodic), we compare it with the previous sequence to determine if a BLE is missing or it is updated to a new value. This helps increase the accuracy of the measurements and estimations. For the cycle scale, we compute the average of every six BLEs that are repeated periodically.

For random-scale long-term experiments that can last hours or days, capturing every MPDU can be costly and incur overheads. To avoid overheads, you can write scripts that periodically send MMs to request BLE, with commands similar to following:

```
1    int6krate -i <ETH_INTERFACE>
```

This command is executed at the station that transmits the data and provides the rates to all the rest of the stations in the network. 🔌

5.4   PLC Link Variability: Comparison with Wi-Fi

We study the spatiotemporal variation of Wi-Fi versus PLC to explore the possibilities of combining the two mediums toward quality of service improvement, coverage extension, and bandwidth aggregation. We then discuss the challenges of hybrid implementations.

We first compare the spatial variation of Wi-Fi and PLC in our testbed, with Wi-Fi and PLC interfaces having similar nominal capacities.[2] This study quantifies the gains that PLC can yield in situations with wireless "blind spots" or bad links and also examines which medium an application should use. We conduct the following experiment: for each pair of stations, we measure the available throughput of both mediums back to back for 5 min, at 100 ms intervals. These experiments are carried out during working hours to emulate a realistic domestic/enterprise environment. We show the average and standard deviation of these measurements (for links with a nonzero throughput for at least one medium).

Let T_W and σ_W be, respectively, the average value and standard deviation of throughput for Wi-Fi (T_P and σ_P, respectively, for PLC). Figure 5.11 shows the results of our experiment. The key findings are as follows.

Connectivity: PLC can yield a better connectivity than Wi-Fi. 100 percent of station pairs that are connected with Wi-Fi are also connected with PLC. In contrast, 81 percent of station pairs that are connected by PLC links, are also connected by Wi-Fi links. At long distance (more than 35 m), there is no wireless connectivity, whereas PLC offers up to 41 Mbps. Thus, PLC can eliminate Wi-Fi blind spots.

Average performance: 52 percent of the station pairs exhibit throughput higher with PLC than with Wi-Fi. PLC can achieve throughput up to 18 times higher than Wi-Fi (40.1 versus 2.2 Mbps). The maximum gain of Wi-Fi versus PLC was similar, that is, 12 times (46.3 versus 3.8 Mbps).

Variability: At short distances (less than 15 m), Wi-Fi usually yields higher throughput, but PLC offers significantly lower variance. Wi-Fi has higher variability, with the maximum standard deviation of throughput being $\sigma_W = 19.2$ Mbps versus $\sigma_P = 3.8$ Mbps for PLC. The vast majority of PLC links yield a σ_P smaller than 4 Mbps.

Conclusion: At long distances, PLC can eliminate wireless blind spots or bad links, yielding gains. At short distances, although Wi-Fi can provide higher throughput, PLC can provide significantly lower variance, which can be beneficial for TCP or applications with demanding, constant-rate requirements, such as high-definition streaming. We explain this difference by the ability of PLC to adapt each carrier to a different modulation scheme, contrary to Wi-Fi (see Section 2.4). PLC reacts more efficiently to bursty errors than Wi-Fi, which has to lower the modulation scheme at all carriers.

---

[2]   We use 802.11n Wi-Fi , with two spatial streams, 20 MHz bandwidth and 400 ns guard interval, yielding a maximum PHY rate of 130 Mbps. We selected a frequency that does not interfere with other wireless networks in our building. The highest PLC data rate is 150 Mbps, hence the interfaces have similar nominal capacities. This is confirmed by the maximum throughputs exhibited by both mediums, shown in Figure 5.11.

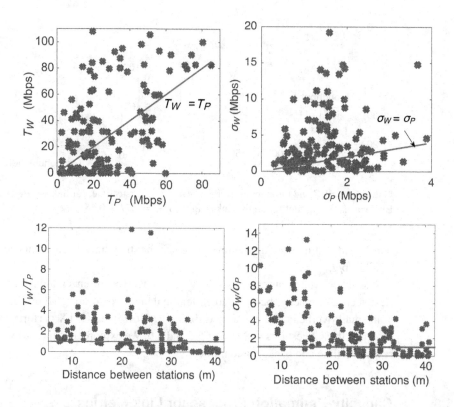

Figure 5.11 Wi-Fi versus PLC performance for all links (top). Spatial variation of the performance ratio between Wi-Fi and PLC (bottom).

5.4.1   Challenges in Hybrid Networks

As we observe in Figure 5.11, although PLC boosts network performance, there are still a few links that perform poorly with both Wi-Fi and PLC. As a result, mesh configurations, hence routing and load balancing algorithms, are needed for seamless connectivity in home or office environments. A challenge for these algorithms is that they have to deal with two different interference graphs with diverse spatiotemporal variation, and that, to fully exploit all mediums, they require accurate link metrics for capacity and loss rates. To this end, unicast probes must be exchanged among the stations. Broadcast packets can be used to estimate metrics of multiple links simultaneously. However, they fail to estimate capacity (see, e.g., [5], [6]). In a network of n stations, unicast probing introduces an $O(n^2)$ overhead that can be significantly reduced by using temporal variation studies of each medium.

A significant challenge is the accuracy of established quality metrics for Wi-Fi, such as the expected transmission count (ETX) or time (ETT) [6], in modern networks, that is, 802.11n/ac/ax. Sheshadri and Koutsonikolas [7] show that due to the MAC/PHY enhancements such as frame aggregation introduced in 802.11n, these metrics perform

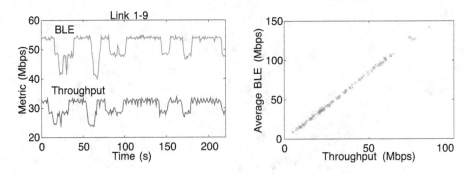

Figure 5.12 BLE and throughput averaged every 1 s, versus time, for link 1-9 (left). Average BLE versus throughput for all the links (right) $BLE = 1.7T - 0.65$.

poorly and that they should be revised, given that they have been evaluated only under 802.11a/b/g.

The above arguments can raise a few questions: How often should the PLC link metrics, such as capacity, be updated in load-balancing or routing algorithms to achieve both small overhead and accurate estimation? How would ETX perform in PLC? We look into these questions in Sections 5.5 and 5.6. We explore link-metrics variation with respect to time and background traffic.

5.5   Capacity-Estimation Process for Link Metrics

We explore a capacity-estimation process for PLC. As mentioned in Section 2.4, stations estimate a tone map, hence also BLE, if and only if they have data to send. Thus, to estimate link metrics, a few unicast probe packets have to be sent. In the previous section, we discuss the rate of the capacity changes given the link quality by sending saturated traffic. Here we examine how capacity can be estimated with a few probe packets, and we explore the size and other features of these packets.

5.5.1   BLE as a Capacity Estimator

First, we show that BLE, which is included in the frame control of every PLC MPDU, can accurately estimate the capacity of PLC links. We repeat saturated tests for our 144 links and with a duration of 4 min. Figure 5.12 presents the measured throughput and BLE for a sample link (1-9) as a function of time. It also presents average BLE versus throughput for all links. For each link, we average BLE over all MPDUs transmitted per link. We observe that BLE is an exact estimation of the actual throughput received by the application. If T is the average throughput, we get $BLE = 1.7T - 0.65$ by fitting a line to the data points. We verified that the residuals are normally distributed.

We next discuss a capacity-estimation technique that uses BLE and probe packets. To conduct a capacity estimation using BLE, a few packets per mains cycle and per estimation interval should be captured, given our temporal variation study in Section 5.3. Here, we investigate an alternative technique that uses MMs to request the instantaneous BLE. The PLC devices provide statistics of the average BLE used over all six tone map slots. Probe packets do not need to be sent at all tone map slots of the mains cycle because, according to the IEEE 1901 standard, the channel-estimation process yields a BLE for all slots when at least one data packet is sent (see Section 2.7).

We now explore whether the number of the probes affects the estimation. To this end, we reset the devices before every run. We perform experiments to estimate the capacity, by sending only a limited number of packets of size 1300 bytes per second (1–200 packets per second).[3] Figure 5.13 shows that the estimated capacity converges to a value that does not depend on the number of packets sent; however, the number of packets sent per second can affect the convergence time of the real estimation. We observe that the channel-estimation algorithm can potentially have a large convergence time to the optimal allocation of bits per symbol for all the carriers, because it needs many samples from many PBs to estimate the error for every frequency, that is, OFDM subcarrier. This convergence time depends on the (vendor-specific) channel-estimation algorithm and on the initial estimation (which was reset in this test).

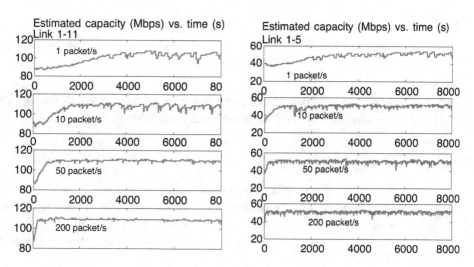

Figure 5.13 Estimated capacity with different numbers of packet probes per second.

To evaluate the convergence time in realistic scenarios, we also perform a test in which we reset the devices at the beginning, but after 2300 s, we pause the probing for approximately 7 min. Figure 5.14 shows the results of the experiments for various links. We observe that the devices maintain the channel-estimation statistics, as the estimated capacity resumes from the previous value before stopping the probing process. Thus, the convergence time of the capacity estimation might not apply in realistic routing implementations and estimation probing for PLC.

To conclude, capacity can be estimated by sending probe packets and measuring BLE in PLC networks. To estimate capacity, given the study in Section 5.3.1, we take into

---

[3] Probe packets can be of any size. PLC always transmits at least a PB (512 B), using padding.

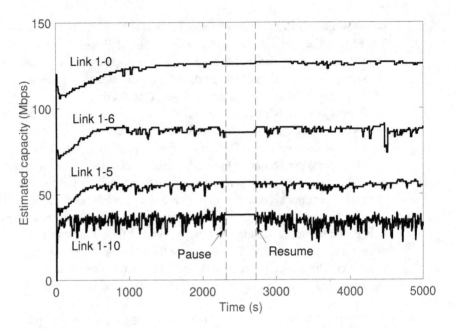

Figure 5.14 Estimated capacity for various links by probing with 20 packets per second. After 2300 s, we stop the probing for 7 min.

account the invariance scale and either compute the average $BLE = \sum_{s=1}^{6} BLE_s/6$ by capturing the PLC MPDUs or request it using MMs.

5.5.2   Size of Probe Packets

We study the appropriate size of probe packets, as large packets can cause communication overheads. We observe that for the special case of sending one probe packet with size less than one PB per second, the estimation might converge to a value smaller than the true one for AV and remain constant with time, independently of the channel conditions.

A representative example of this behavior is shown in Figure 5.15, where AV capacity converges to approximately 89 Mbps when sending only one packet per second with size less than one physical block (520 bytes including PB header of 8 bytes). After this convergence, the estimated capacity remains constant. A simple computation shows that the rate required to transmit one PB in one OFDM symbol is $R_{1sym} = (520 \times 8)/(T_{sym} + GI) \approx 85.7$ Mbps with AV, given the symbol duration $T_{sym} = 40.96$ μs and guard interval 7.56 μs. When sending packets smaller than one PB, the rate converges close to R_{1sym} for all six slots of the mains cycle, because increasing the rate does not reduce the transmission time (it is not possible to transmit less than one OFDM symbol) while decreasing the probability of error (higher rates yield less robust modulation schemes).[4] Hence, we find that to estimate the capacity of a link by sending only one probe packet

---

[4] This behaviour is vendor specific and might not be observed for all PLC devices.

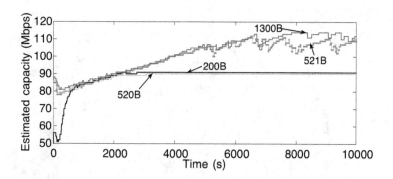

Figure 5.15 Estimated capacity for link 11-6: one probe packet per second, and various sizes.

per second, it is crucial to send packets larger than one PB or one OFDM symbol. To estimate rates higher than $R_{nsym} = (n \times 520 \times 8)/T_{sym} \approx n \times 85.7$ Mbps, it can be important to send at least n PB per probe, possibly with several MSDUs.

5.5.3 Priority Class of Probe and Control Packets

While running experiments with different priority classes (see Section 3.5), we notice an abnormal behavior with priority class CA3 and UDP traffic: there are periodic through-put drops that are not due to BLE for all the links. An example of such behavior is depicted in Figure 5.16. BLE is an accurate estimation of throughput for the same link and class CA1 in Figure 5.12. However, for CA3 class, we observe throughput drops pe-riodically. By using `tcpdump`, we discover that the throughput drops are due to the fact that ARP packets have a priority (CA1) lower than the data packets (CA3), thus they do not contend for the medium, unless the CA3 priority queue is empty. As explained in Section 3.5, stations must defer their transmission in case of a higher priority class in the network based on the priority-resolution process. While ARP packets wait for the CA3 queue to be emptied, new data packets are discarded, as the ARP cache does not contain any information about the IP address of the destination.

We observe that all links exhibit the same behavior with the CA3 class UDP traffic. Figure 5.16 also presents a histogram of the interarrival times of throughput drops of all links of our testbed. We find that the throughput drops are synced with the ARP requests that are sent when entries in the ARP cache expire after an interval called the base reachable time. Let b be the base reachable time the Linux system uses [8]. Typically ARP entries expire with a period chosen uniformly at random over the range $[b/2, 3b/2]$; b defaults to 30 s for the Linux boards used in these tests. We observe that the values of the histogram in Figure 5.16 are in this interval. We do not observe the same phenomenon with TCP traffic, because the TCP-ACKs sent by the destination station are used to update the ARP table by our boards. This is an important feature of PLC implementations that can harm the QoS of UDP-based applications that are tagged with the CA2 or CA3 classes.

Given these experimental results, in hybrid network implementations, management or

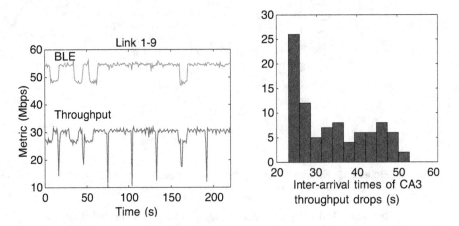

Figure 5.16 BLE and throughput of CA3 class averaged over 1 s intervals, versus time (left). Histogram of interarrival times of CA3 throughput drops from ninety links (right).

control messages should be tagged as high priority. Control messages or probe packets, such as ARP, should be tagged with the same or higher priority as the data frames.

5.5.4   Validation of Capacity Estimation in Hybrid Single Hop

To evaluate the capacity-estimation method, we design a simple load-balancing algorithm that aggregates bandwidth between Wi-Fi and PLC and operates between the IP and MAC layers. To implement our algorithm, we use the Click Modular Router and elements that we describe in Section 4.8 [9]. Figure 5.17 shows the Click elements used. The "LoadBalancer" forwards each IP packet to one of the mediums (interfaces eth2 or wifi0) with a probability proportional to the capacity of the medium. For PLC, the capacity estimation uses MMs sent periodically from "LoadBalancer," and for Wi-Fi, it reads PHY rate statistics from the driver (by default written in one of the file systems).[5] The "LoadBalancer" receives the PLC MM replies at input 1 (classified as EtherType 0x88e1 by a Classifier element) and IP packets at input 0. At the destination, we reorder the IP packets using a simple algorithm that checks the identification sequence of the IP header. The element "PacketReorder" implements this algorithm on a per-flow (UDP or TCP) basis. A flow is defined by using the 5 tuple of source and destination IP, source and destination port, and protocol. This example is implemented for single-hop load balancing.

To estimate the capacities, we probe links with one packet per second. The capacity for PLC is estimated every second using BLE, that is, averaged over the six tone map slots of the invariance scale, whereas for Wi-Fi, the capacity is estimated using

---

[5] For both Wi-Fi and PLC, it is possible to estimate the link capacity by setting the interface in sniffer mode (with faifa for PLC, with a monitor interface for Wi-Fi), and by reading for every packet the BLE field for PLC and the MCS field for Wi-Fi. This might, however, cause a high CPU utilization.

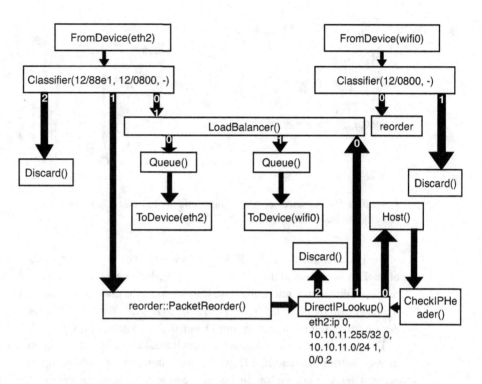

Figure 5.17 A simple Click! design that incorporates a load balancer with capacity-estimation techniques "LoadBalancer."

the modulation and coding scheme index (MCS).[6] MCS is averaged over the transmissions (data and probes) every second using a moving average. As we observe in Section 5.4, Wi-Fi can have higher variability than PLC, and hence we use a moving average of Wi-Fi measurements to reduce the errors due to Wi-Fi high MCS standard deviation. The load-balancing algorithm takes into account our temporal-variation study on PLC: in Section 5.3.1, we find that the PLC channel quality is periodic, with every packet using a different BLE. Because an accurate synchronization at this timescale is challenging for algorithms operating above the MAC layer (such as in IEEE 1905.1 standard), the capacity of PLC in hybrid networks can be estimated by averaging over the invariance scale.

In Figure 5.18, we first present the throughput of experiments on one link. We run four experiments back to back: using only one of the interfaces (Wi-Fi, PLC) in two; using both interfaces and our load-balancing algorithm (Hybrid) in one; and using both interfaces and a round robin scheduler for the packets in the last one. We observe that by using simple load-balancing and reordering algorithms, and the aforementioned capacity-estimation technique, we can achieve a throughput that is very close to the sum of the capacities of both mediums. In contrast, the throughput of a round robin scheduler, which has no information on capacity, is limited to twice the minimum capacity

---

[6] Given MCS index, we retrieve the capacity from IEEE 802.11n standard tables [10].

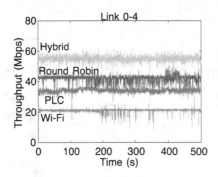 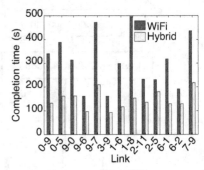

Figure 5.18 Performance boost by using hybrid Wi-Fi/PLC and load-balancing and capacity-estimation techniques.

of the two mediums (i.e., Wi-Fi in this example), because it assigns the same number of packets to each medium and the slowest medium becomes a bottleneck. To evaluate our algorithm across our testbed, we also compare the completion times of a 600 Mbyte file download by using (1) only Wi-Fi and (2) Hybrid. We observe in the same figure, a drastic decrease in completion times when using both mediums.

The above tests show that, to exploit each medium to the fullest extent, accurate link-quality metrics are required. However, the link metrics should be updated to take into account delay or contention. In the next section, we investigate another link metric, that is, the expected number of retransmissions and the performance of link metrics with respect to background traffic.

⚏ Experimental Guideline In this section, we use multiple tools to estimate the capacity and send traffic. First, you can modify traffic intensity and packet sizes using standard Linux tools (`iperf` or `ping`). Second, you can use `Click!` or other software to generate your own IP packets (see Section 4.8 for the detailed code). We change the PLC priority class from CA0 to CA3 by using the type of service field in the IP header. By default, CA1 class is used for packet transmission. To generate CA3 class MPDUs, use the `-S` option of `iperf` or `-Q` option for `ping`. The ToS value of 192 generates CA3 class in the devices we have tested. However, other PLC devices might be configured differently and might require setting the QoS VLAN priority of the Ethernet header. For many PLC devices in the market, the mapping of QoS to CA class in PLC is configurable through the vendor's software. ⚏

5.6   Retransmitting in PLC Channels

Capacity is a good metric for link quality. However, it does not take into account interference, which is very important for selecting links in shaded mediums like Wi-Fi and PLC. Moreover, another metric could be useful for delay-sensitive applications that do not saturate the medium and do not need high capacity, but have low delay requirements.

Delay can be affected by retransmissions either due to bursty PHY errors or to MAC contention in addition to reduced PHY rates. The PB error rate PB_{err} introduced in Section 2.7 (or packet errors of IEEE 1905 [11]) is correlated with retransmissions, and it can be increased due to PHY errors or MAC collisions. We explore the mechanism of retransmissions in PLC networks. We first study the expected transmission count (ETX). Numerous works (e.g., [5, 6]) study this metric (or its variations) in Wi-Fi networks by sending broadcast probes. We examine how ETX performs in PLC and the relationship between broadcast and unicast probing. After studying retransmissions due to errors, we evaluate the sensitivity of link metrics to background traffic.

5.6.1 Retransmission due to Errors

We first explore how ETX would perform in PLC by sending broadcast packets. Because broadcast packets are transmitted with the most robust modulation and are acknowledged by some proxy station [4], we expect that this method yields very low loss rates.

For the purpose of this study, we set each station in turn to broadcast 1500-byte probe-packets (one every 100 ms) for 500 s. The rest of the stations count the missed packets by using an identification in our packet header. We repeat the test for all stations of the testbed during night and working hours (day). Figure 5.19 shows the loss rate as a function of throughput and PB_{err} from our tests for all station pairs. The loss rate of broadcast packets in PLC is a very noisy metric for the following reasons:

(1) A wide range of links with diverse qualities have very low loss rates ($\sim 10^{-4}$), and some links even have 0 loss rates. By observing high loss rates, e.g., larger than 10^{-1}, ETX can classify bad links in PLC; but nothing can be conjectured for link quality from low loss rates.

(2) There is no obvious difference between experiments during the day, when the channel is worse, and night. A few bad links have poor loss rates during the day, but at the same time, a few average links yield even lower loss rates.

(3) As PLC adapts the modulation scheme to channel conditions when data is transmitted, broadcast packets – sent at the most robust modulation scheme – cannot reflect the real link quality, as the link will use different BLE for data transmission. Given the low loss rates of a wide range of links, ETX appears to be 0 at short timescales, which provides no or misleading information on link quality.

Owing to the above observations, we further explore the mechanism of retransmissions with respect to link quality with unicast traffic. We now measure retransmissions of PBs by sending unicast, low data-rate traffic, that is, 150 kbps, and by capturing the PLC frame headers. Under this scenario, an Ethernet packet of 1500 bytes is sent approximately every 75 ms. The test has a duration of 5 min per link.

As we discuss above, broadcast packets might be missed by some stations when channel conditions are bad, because they are not retransmitted as soon as a proxy station acknowledges them. In contrast, unicast packets are being retransmitted until the receiver acknowledges them, hence they are always received unless there is very high delay due to queuing or packet retransmissions and the packets are dropped (timeout for MPDUs

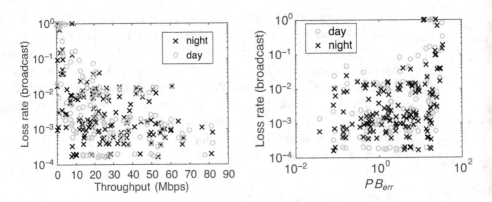

Figure 5.19 Loss rate for broadcast packets versus link throughput and PB_{err} for all station pairs.

is in order of hundreds milliseconds from Table 3.2). For this reason, we look at the frame header (SoF) to study retransmissions. Because there is no indication whether the frame is retransmitted in the PLC SoF, we use the arrival timestamp of the frame to characterize it as a retransmission or new transmission (if the frame arrives within an interval of less than 10 ms compared to the previous frame, then it is a retransmission). We also measure PB_{err} every 500 ms.

We conduct the experiment described above for all the links of our testbed. We compute the unicast ETX (U-ETX) for all the links of the testbed. We count the total number of retransmissions for a packet of 1500 bytes, which produces three PBs. A retransmission occurs if at least one of these PBs is received with errors. Figure 5.20 presents U-ETX as a function of average BLE (with links sorted in increasing BLE order) and PB_{err}. U-ETX is measured by averaging the number of PLC retransmissions for all packets transmitted during the experiment. We also plot error bars with the standard deviation of the transmission count. We observe that link quality is negatively correlated with link variability, a conclusion made also when exploring BLE in Section 5.3.2. The higher the U-ETX is, the higher the standard deviation of transmission count is. Links with high BLE are very likely to guarantee low delays, as U-ETX does not vary a lot. U-ETX and the averaged PB_{err} are highly correlated, with almost a linear relationship.

5.6.2   Retransmission due to Contention

To explore the sensitivity of link metrics to background traffic and to examine how interference can be considered in link metrics, we now experiment with two contending flows. We set a link to send unicast traffic at 150 kbps, as in the previous subsection, thus emulating probe packets. After 200 s, we activate a second link sending "background" traffic at various rates. We measure both BLE and PB_{err}. In these experiments, we observe that BLE is insensitive to low-data-rate background traffic for all pairs of links. However, BLE appears to be affected by high-data-rate background traffic on a few pairs of links. So far, we have not found any correlation between these pairs of links. We explain this phenomenon with the "capture effect," where the best link decodes a

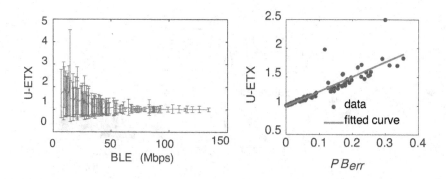

Figure 5.20 U-ETX versus BLE and U-ETX versus PB error rate.

few PBs even during a collision due to very good channel conditions, thus resulting in high PB_{err}. In this case, the channel-estimation algorithm cannot distinguish between errors due to PHY layer and errors due to collisions, hence it decreases BLE. Figure 5.21 presents two representative examples of link pairs for which BLE is sensitive and non-sensitive to high data-rate background traffic. PB error rate increases drastically in link 6-11, which is sensitive to background traffic.

To address the sensitivity of our link metrics to high-data-rate background traffic, we take advantage of the frame aggregation procedure of the MAC layer, which is described in Section 3.1. We observe that transmitting a few PBs per 75 ms (150 kbps rate) yields a sensitivity of metrics to background traffic. However, when two saturated flows are activated, we never notice an effect on BLE. Owing to frame aggregation, packets from different saturated flows have approximately the same frame length (i.e., maximum), and when they collide, the channel estimation algorithm works more efficiently than when short probe packets collide with long ones. To emulate the long frame lengths of saturated traffic, we send bursts of twenty packets such that the traffic rate per second (i.e., the overhead) is kept the same (150 kbps). In Figure 5.22, we show another link for which BLE is sensitive to background traffic, and the results of the proposed solution. By sending bursts of probe packets, BLE is no longer affected by background traffic. This shows that by taking into account the frame aggregation process,[7] we can design capacity metrics that are insensitive to background traffic.

To conclude, we have studied the mechanism of retransmissions in PLC. Although broadcast probe packets yield significantly less overhead in link-quality estimation, they do not provide accurate estimations. In contrast, unicast probe packets reflect the real link quality, but by producing more overhead. We observe that PB_{err} can be used to estimate U-ETX and to indicate interference in PLC. However, estimating the amount of interference is challenging and should be further investigated.

📋 **Experimental Guideline** In this section, we measure multiple error rates. First, we

---

[7] Depending on the PLC technology, these bursts can be transmitted such that only one PLC MPDU is generated, hence without large MAC overhead.

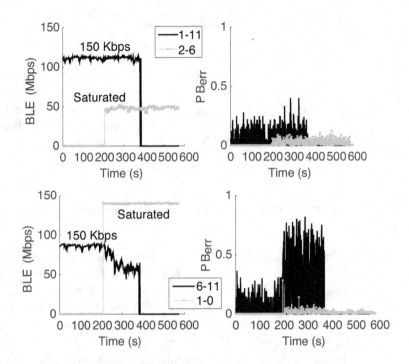

Figure 5.21 Link metrics of two sets of contending links with low data rate and saturated traffic. The link metrics on the bottom plot are sensitive to background traffic.

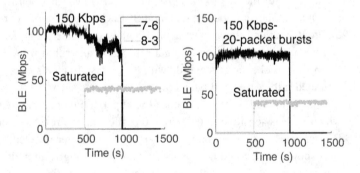

Figure 5.22 Addressing the link metric sensitivity to background traffic by sending bursts of probes.

measure the error rate of broadcast packets. We write a Click! script where we add our own header on broadcast packets, and we measure the missed packets based on an identification field on our header. We repeat the experiment for all stations in the network by broadcasting thousands of packets and by using a Click! script at every PLC device to measure the packet losses using packet identification. Second, we measure the unicast loss rate by enabling the sniffing mode of the PLC device (see Sections 5.3 and 4.7.3) and sending less frequent data packets (e.g., once in 10 ms) to avoid frame

Table 5.1 Guidelines for PLC link metric estimation

Policy	Guideline/explanation	Section
Metrics	BLE and PB_{err}, defined by IEEE 1901.	5.5, 5.6.1
Unicast probing only	Broadcast probing cannot be used, as it does not give sufficient information on link quality.	5.6.1
Shortest timescale	BLE should be averaged over the mains cycle.	5.3.1
Size of probes	Larger than one PB (or one OFDM symbol) to avoid inaccurate convergence of rate adaptation algorithm.	5.5.2
Frequency of probes	Should be adapted to link quality for lower overhead.	5.3.2, 5.3.3
Burstiness of probes	Can tackle a potential inaccurate convergence of the channel estimation algorithm or the sensitivity of link metrics to background traffic.	5.5.2, 5.6.2
Asymmetry in probing	There is both spatial- and temporal-variation asymmetry in PLC links. This could affect bidirectional traffic, such as TCP, that requires routing in both directions.	5.2, 5.3.2
Priority of control traffic	Control messages should be tagged at the same priority as the data to avoid performance degradation.	5.5.3

aggregation. By studying the captured MPDU patterns, we can infer whether we have a retransmission: since we send packets only every 10 ms, every packet seen in between these periodic intervals and close to the first original transmission, is counted as a retransmission. Finally, we measure PB_{err} by sending MMs and using commands such as int6kstat, as described in Sections 4.6.2 and 5.2.

5.7 Link-Metric Guidelines

One of the main discussions in this chapter involves the systematic guidelines for link metric estimation in PLC and hybrid networks. In this section, we summarize these guidelines, which might help in the future implementation of heterogeneous networks that include PLC. Table 5.1 outlines our guidelines for efficient link-metric estimation with PLC, given our experimental study in this chapter. We present the link metrics proposed in this chapter and instructions on how to perform their estimation with high accuracy and low overhead. In addition, we refer the reader to the relevant sections in this chapter.

5.8   Summary

In this chapter, we have explored experimentally the end-to-end performance of isolated PLC links in an enterprise environment, focusing on spatial and temporal variations of capacity. We have found that PLC can provide very good connectivity for stations connected to the same distribution panel, forming an almost fully connected graph. However, the exact link quality is challenging to predict given station location, due to the severe asymmetry PLC links often exhibit. Concerning the temporal variation, we have found that it occurs in three timescales because of the diverse types of noise in PLC channels and because of human behavior with respect to electrical activity. Yet, PLC links show low variability, hence they can serve bandwidth-hungry applications.

We have also studied PLC link metrics and their variation with respect to space, time, and background traffic. Similar metrics have been investigated for Wi-Fi and have been required by the recent standardization of hybrid networks. We have proposed practical guidelines on efficient metric estimation in hybrid implementations. We have observed that there is a high correlation between link quality and its variability, which has a direct impact on probing overhead and accurate estimations.

References

[1] R. Murty, J. Padhye, R. Chandra, A. R. Chowdhury, and M. Welsh, "Characterizing the end-to-end performance of indoor powerline networks," Harvard University and Microsoft Research, Tech. Rep., 2008.

[2] S. Sancha, F. Cañete, L. Diez, and J. T. Entrambasaguas, "A channel simulator for indoor power-line communications," in *2007 IEEE International Symposium on Power Line Communications and Its Applications (ISPLC)*, pp. 104–109.

[3] S. Guzelgöz, H. B. Çelebi, T. Güzel, H. Arslan, and M. K. Mihçak, "Time frequency analysis of noise generated by electrical loads in PLC," in *Proceedings of the 2010 17th IEEE International Conference on Telcommunications (ICT)*, pp. 864–871.

[4] *IEEE Standard for Broadband over Power Line Networks: Medium Access Control and Physical Layer Specifications*, IEEE Standard 1901-2010.

[5] S. M. Das, H. Pucha, K. Papagiannaki, and Y. C. Hu, "Studying wireless routing link metric dynamics," in *ACM Internet Measurement Conference (IMC) 2007*, pp. 327–332.

[6] R. Draves, J. Padhye, and B. Zill, "Routing in multi-radio, multi-hop wireless mesh networks," in *ACM 10th Annual International Conference on Mobile Computing and Networking (MobiCom)*, 2004, pp. 114–128.

[7] R. K. Sheshadri and D. Koutsonikolas, "Comparison of routing metrics in 802.11n wireless mesh networks," in *IEEE INFOCOM International Conference on Computer Communications*, 2013, pp. 1869–1877.

[8] C. Benvenuti, *Understanding Linux Network Internals*. O'Reilly Media, 2006.

[9] E. Kohler, R. Morris, B. Chen, J. Jannotti, and M. F. Kaashoek, "The Click modular router," *ACM Transactions on Computer Systems*, vol. 18, no. 3, pp. 263–297, 2000.

[10] *Enhancements for Higher Throughput*, IEEE 802.11n-2009-Amendment 5.

[11] *IEEE Standard for a Convergent Digital Home Network for Heterogeneous Technologies*, IEEE Standard 1905.1-2013.

6 MAC-Layer Performance Evaluation

6.1 Introduction

In the previous chapter, we mainly discussed the capacity and variability of isolated PLC links. We now analyze PLC MAC performance under multiuser scenarios, where multiple stations transmit data simultaneously. As we have seen in Sections 2.2.3 and 3.5, PLC is a shared medium and it uses a CSMA/CA protocol in order to avoid and react to collisions among stations. The PLC CSMA/CA includes an additional variable compared to the one of Wi-Fi, which is called the deferral counter. Despite the CSMA collision avoidance, the stations can still collide, which can waste airtime and decrease throughput. We discuss how the total throughput scales with the number of contending stations with analytical models, Wi-Fi/PLC testbeds, and simulations.

In Section 6.2, we present a mathematical model to analyze the throughput of the PLC CSMA/CA. The model relies on the assumption that the station MAC processes are independent, which is often called the decoupling assumption in the literature. We compare the model to simulations and experiments with commercial PLC chipsets in Section 6.3. Finally, we show a PHY/MAC cross-layer evaluation in Section 6.4, to understand if low-PHY-rate stations can limit the throughput of high-rate ones. We discuss whether cross-layer performance anomalies found for early 802.11a (Wi-Fi) networks manifest for PLC. This performance anomaly depends on PHY rates of the stations and the frame aggregation process described in Section 3.3.

In Section 6.5, we discuss the short-term fairness of the PLC CSMA/CA and show that high throughput of the PLC MAC layer can come at a cost of short-term fairness. We present another model for fairness and show that the default PLC CSMA/CA priority class can be short-term unfair, with two metrics analytically and in simulation. We show the impact of this unfairness to realistic network delays, and in Section 6.6, we illustrate the trade-off of PLC MAC layer between throughput and fairness.

The deferral counter in PLC CSMA/CA can result in a certain level of unfairness and coupling between the stations depending on its values. This coupling can reduce the accuracy of models based on the decoupling assumption, as we discuss in Section 6.7. This assumption was originally proposed in Bianchi's [1] analysis of the Wi-Fi CSMA/CA and has been used in all works that have analyzed the PLC CSMA/CA so far [2, 3] and the throughput model presented in Section 6.2. In Section 6.8, we discuss how modeling accuracy can be improved by avoiding this hypothesis, and we introduce a second and more complex model.

6.2   The Decoupling Assumption Model

6.2.1   Modeling Assumptions

We use the following networking assumptions, which have also been used for modeling Wi-Fi and PLC CSMA/CA [1, 2, 4, 5] to make the analysis tractable.

Perfect sensing of transmissions: We have N stations hearing each other, hence a single collision domain. There are no hidden terminals.

Saturated queues: All stations always have a packet to transmit in their queue.

Perfect PLC channel: There are no packet losses or errors due to the PHY layer. Thus transmission failures are only due to collisions. All stations use the same PHY rate, and all frames have equal durations. As a result, this model can be used for different data rate HomePlug specifications, as they all use the same CSMA/CA process.

Infinite retry limit: All stations have an infinite retry limit under collisions. They never discard a packet until it is successfully transmitted. Contrary to the IEEE 802.11 standard, IEEE 1901 does not specify a retry limit. However, there is a timeout on the frame transmission that is vendor specific, as we discuss in Section 3.4. For example, the commercial devices tested in Section 6.3.1 have a timeout of 2.5 s for CA1 priority, which is very large compared to the maximum frame duration (2.5 ms according to the standard [6]). Therefore the infinite retry limit assumption is reasonable.

Homogeneous network: The IEEE 1901 standard introduces four different priority classes (see Section 3.4) and specifies that only the stations whose frames belong to the highest contending priority class run the backoff process. Therefore, we consider a practical scenario in which all contending stations use the same set of parameters, that is, those of the highest priority class present in the network.

The modeling assumption of the PLC MAC model is called the decoupling assumption (see, e.g., [1, 4]). According to it, the backoff process of a station is independent of the aggregate attempt process of the other $N - 1$ stations, which results in the following approximations:

- Given a tagged station, the probability that at least one of the other stations transmits at any time slot is fixed, equal to γ, from which (1) transmission attempts experience a fixed collision probability γ and (2) a station with $BC \neq 0$ senses the medium busy at any time slot with a fixed probability γ. We can compute the probability that a station transmits in a randomly chosen time slot, which we denote by τ, as a function of γ.
- The transmission attempts of different stations are independent, with the transmission probability of a station given by the average attempt rate τ. Hence, we can express γ as a function of τ, leading to a fixed-point equation that can be solved to obtain the γ, τ values.

6.2.2   Mathematical Analysis

Recall that the stations transmit at randomized intervals by using a uniformly distributed random variable, called the backoff counter BC. The backoff counter has a maximum value called the contention window (CW). A station starts any new transmission attempt at backoff stage 0, and in case of a collision or sensing the medium busy, it can move to higher backoff stages with larger contention-window values. It decrements the backoff counter at every idle slot until it reaches $BC = 0$ when the station transmits (see Section 3.5). We base the analysis on the renewal-reward theorem. A renewal process is a random point process for which the interarrival times form an independent and identically distributed (i.i.d.) sequence. In the PLC MAC case, the successful transmission occurrences form an i.i.d. sequence, as their backoff/deferral counters and backoff stage are reset after every successful transmission, and the exact same process is repeated for the next transmission. Figure 6.1 shows an example of this process. X_j is the total backoff time for the jth frame, given by $X_j = \sum_{i=0}^{R_j-1} B_j^{(i)}$, where $B_j^{(i)}$ is the backoff counter value chosen at backoff stage i for the jth frame. R_j is the number of transmission attempts for frame j. X_j and R_j are i.i.d. sequences, and the number of attempts R_j can be viewed as a reward associated with the renewal cycle length of X_j.

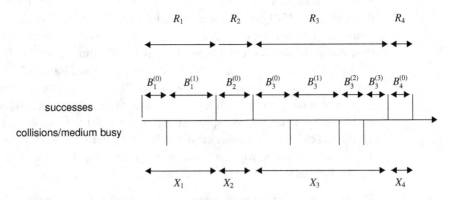

Figure 6.1 An example of the evolution of the backoffs of a station over time after removing channel (transmission durations) activity in a PLC network. Each attempted transmission starts a new backoff cycle.

The renewal-reward analysis needs the expected number of time slots spent by a successful transmission, hence also the time slots a station spends at any backoff stage i. In CSMA/CA analysis, we typically assume that a time slot can either be idle or contain a transmission [1, 2, 4, 5]. Let k be the value of BC drawn uniformly at random in $\{0, \ldots, CW_i - 1\}$, when the station enters backoff stage i. If the station is a Wi-Fi one, the station leaves the backoff stage when (and only when) it attempts a transmission; hence, the station stays in the backoff stage exactly $k + 1$ time slots, until BC expires and it attempts a transmission. However, a PLC station might leave backoff stage i either because of a transmission attempt, when BC expires (like in Wi-Fi), or because it has sensed the medium busy $d_i + 1$ times (d_i is the maximum value of deferral counter at backoff stage i, as defined in Table 3.4) before BC has expired. In the latter case, the station spends a number of slots at backoff stage i equal to j, if it senses the medium busy for the $(d_i + 1)$th time in the jth slot, with $d_i + 1 \le j \le k$.

We let bc_i be the expected number of time slots spent by a station at backoff stage i. To compute bc_i, we need to evaluate the probability of the events that (1) a station attempts a transmission or (2) senses the medium busy d_i+1 times within the k slots of the backoff counter (i.e., before BC expires). Let b be the random variable describing the number of busy slots within the k backoff slots. Because of the decoupling assumption that yields a constant medium-busy probability across slots, b follows the binomial distribution $\text{Bin}(k, \gamma)$, where γ is the probability of sensing the medium busy. Now, let x_k^i be the probability that a station at backoff stage i jumps to the next stage $i + 1$ in k or fewer time slots due to event (2). Then, we have

$$x_k^i = \mathbb{P}(b > d_i) = \sum_{j=d_i+1}^{k} \binom{k}{j}\gamma^j(1 - \gamma)^{k-j}, \tag{6.1}$$

where d_i is given by Table 3.4.

We can compute bc_i as a function of γ via x_k^i. We have two events that can happen in k slots. First, if $k > d_i$, then event (1) occurs with probability $1 - x_k^i$, and event (2) occurs with probability x_k^i. For event (i), the station spends $k + 1$ slots in stage i. For event (2), the station spends j slots in backoff stage i with probability $x_j^i - x_{j-1}^i$ for $d_i + 1 \leq j \leq k$ (because $x_j^i - x_{j-1}^i$ is the probability that (2) happens exactly at slot j). Second, if $k \leq d_i$, then event (2) cannot happen. Thus, the backoff counter expires, event (1) always takes place, and the station spends $k + 1$ time slots in stage i. By considering all the above cases, bc_i is computed as

$$bc_i = \frac{1}{CW_i} \sum_{k=d_i+1}^{CW_i-1} \left[(k + 1)(1 - x_k^i) + \sum_{j=d_i+1}^{k} j(x_j^i - x_{j-1}^i) \right] + \frac{(d_i + 1)(d_i + 2)}{2CW_i}, \tag{6.2}$$

where x_k^i is given by Equation (6.1) and d_i is given by Table 3.4.

We next compute the probability that a station at backoff stage i ends this stage by attempting a transmission, which we denote by t_i, and the probability that such a backoff stage ends with a successful transmission s_i. Similarly to (6.2), t_i is computed as

$$t_i = \sum_{k=d_i+1}^{CW_i-1} \frac{1}{CW_i}(1 - x_k^i) + \frac{d_i + 1}{CW_i}. \tag{6.3}$$

From the above, we have

$$s_i = (1 - \gamma)t_i, \tag{6.4}$$

since γ is the probability of collision at any random slot.

Based on the above expressions for bc_i, t_i, and s_i, we evaluate the average attempt rate of a station, τ. Let R be the random variable describing the number of transmission attempts experienced by a successfully transmitted frame. Similarly, let X be the random variable for the total number of slots spent in backoff for a successfully transmitted frame. R and X form an i.i.d sequence for consecutive successful transmissions. Hence, from the renewal-reward theorem (R being the reward and X the renewal lifetimes [4]), the average attempt rate is given by

$$\tau = \frac{\mathbb{E}[R]}{\mathbb{E}[X]}. \tag{6.5}$$

We compute $\mathbb{E}[R]$ and $\mathbb{E}[X]$ with the following two lemmas (see proofs in Appendix A).

LEMMA 6.1. *The expected number of slots spent in backoff per successfully transmitted frame is*

$$\mathbb{E}[X] = \sum_{i=0}^{m-2} bc_i \prod_{j=0}^{i-1}(1 - s_j) + \prod_{i=0}^{m-2}(1 - s_i)\frac{bc_{m-1}}{s_{m-1}}. \tag{6.6}$$

LEMMA 6.2. *The expected number of transmission attempts per successfully transmitted frame is*

$$\mathbb{E}[R] = \sum_{i=0}^{m-2} t_i \prod_{j=0}^{i-1}(1 - s_j) + \prod_{i=0}^{m-2}(1 - s_i)\frac{t_{m-1}}{s_{m-1}} = \frac{1}{1-\gamma}. \tag{6.7}$$

We can now form a fixed-point equation on γ. From the decoupling assumption, the probability γ that at least one other station transmits can be expressed as a function of τ:

$$\gamma = \Gamma(\tau) = 1 - (1 - \tau)^{N-1}. \tag{6.8}$$

In addition, $\tau = \mathbb{E}[R]/\mathbb{E}[X]$ can be expressed as a function of γ from Lemmas 6.1 and 6.2. Let this function be $G(\gamma)$. Then, we have

$$G(\gamma) = \frac{1}{1-\gamma} \sum_{i=0}^{m-2} bc_i \prod_{j=0}^{i-1}(1 - s_j) + \prod_{i=0}^{m-2}(1 - s_i)\frac{bc_{m-1}}{s_{m-1}}, \tag{6.9}$$

where bc_i is given by Equation (6.2) and s_i is computed in Equation (6.4). The composition of the functions $\tau = G(\gamma)$ and $\gamma = \Gamma(\tau)$ yields a fixed-point equation for the collision probability:

$$\gamma = \Gamma(G(\gamma)). \tag{6.10}$$

By solving the above fixed-point equation, we can compute the value γ of a PLC network with N stations and any configuration $CW_i, d_i, 0 \le i \le m - 1$. Figure 6.2 shows the solution of Equation (6.10) graphically for the default priority class CA1 of IEEE 1901 and for different numbers of stations. The solution is the intersection of functions $\Gamma(G(\gamma))$ and $f(\gamma) = \gamma$. Equation (6.10) can be solved by standard nonlinear equation solvers (e.g., in Matlab or python).

🛈 Further Reading   Theorem 6.1 below proves the uniqueness of the solution to fixed-point Equation (6.10), under the condition that the transmission probability is decreasing with the backoff stage i. The theorem, with proof in Appendix A, provides examples of configurations satisfying this condition.

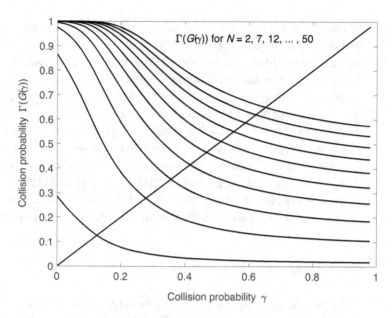

Figure 6.2 The solution of fixed-point equation (6.10) for the default values of the contention window for priority CA1 of IEEE 1901 CSMA/CA protocol.

THEOREM 6.1. $\Gamma(G(\gamma)) : [0, 1] \to [0, 1]$ *has a unique fixed point if the following condition is satisfied for* $0 \le i \le m - 2$:

$$\begin{cases} CW_{i+1} \ge CW_i & \text{if } d_{i+1} = d_i, \\ CW_{i+1} = CW_i & \text{if } d_{i+1} < d_i, \\ CW_{i+1} \ge 2CW_i - d_i - 1 & \text{otherwise.} \end{cases} \tag{6.11}$$

6.2.3 Throughput Evaluation under the Decoupling Assumption Model

The model in the previous section can be used to obtain the throughput of the PLC network under saturated conditions. After we compute the collision probability γ from fixed-point Equation (6.10), we have τ by using Equation (6.8). We can also compute p_s and p_e, respectively, the probability that a slot contains a successful transmission or that it is empty, from $p_s = N\tau(1 - \tau)^{N-1}$ and $p_e = (1 - \tau)^N$. We can further compute the probability that a slot contains a collision as $p_c = 1 - p_e - p_s$. The empty, collision, and successful transmission slots can have different durations. We now have enough information to compute the normalized throughput S of the network as

$$S = \frac{p_s D}{p_s T_s + p_c T_c + p_e \sigma}, \tag{6.12}$$

where D is the frame duration including data, T_s is the duration of a successful transmission, T_c is the duration of a collision, and σ is the time slot duration. T_s and T_c contain various MAC overheads in addition to the data duration D, such as preambles, acknowledgments, and interframe spaces (see Section 3.5.1).

When stations are saturated, frame durations are the same for all stations independently of their rates, because of the frame aggregation process described in Section 3.3 and of the maximum frame duration imposed by the standard. Thus, to compute actual throughput in bps, we have to replace the frame duration D with the number of data bytes per PLC frame in Equation (6.12). The model provides a performance evaluation that is independent of the PLC PHY technology and PHY rates, and that can be applied to any HomePlug specification described in Section 2.3. Therefore, in the rest of this chapter, we generally use the normalized throughput given by Equation (6.12).

6.3   Performance Evaluation

We now compare the accuracy of the model via simulation and experiments with a testbed of Wi-Fi and PLC devices. Then, we explore the PLC multiuser performance over a wide range of configurations.

6.3.1   Experimental Comparison with Simulator and Model

We use simulations to evaluate PLC multiuser performance. We have a `Matlab` simulator that implements the full CSMA/CA mechanism. The simulator is time slotted; it emulates the backoff procedures of the PLC and assumes perfect sensing and error-free channel. The simulator can be built in a single thread since the stations are synchronized, and we assume no hidden terminals. For the extension of the simulations to multiple collision domains, we recommend a discrete-event simulator. We validate the accuracy of the `Matlab` simulator and the model presented in the previous section with experimental results from a HomePlug AV testbed.

For this evaluation, we use a testbed of eight stations with a HomePlug AV interface (a miniPCI card with Intellon INT6300 chip) and a 802.11n interface (a miniPCI card Atheros DNMA-92). The stations are Alix boards running the OpenWrt Linux distribution [7]. In this setup, all PLC devices are plugged in the same extension cord, so that we have ideal channel conditions between the stations and minimum PHY layer errors. To reproduce MAC layer studies and investigate how the protocol scales with the number of contending stations, it is important to ensure ideal PHY conditions and that all stations have similar channel qualities. For instance, if stations have a large difference in channel quality (signal-to-noise ratio), then the station with the best channel might be able to decode the frame even under collision and to monopolize the channel yielding unfairness and wrong throughput/collision estimates. This phenomenon is called the capture effect in the literature.

In these experiments, we have N stations send UDP traffic (at a rate higher than the link capacities) to the same nontransmitting station using `iperf`. We run experiments for $1 \leq N \leq 7$, and we measure the collision probability. Figure 6.3 shows average collision probability from 10 testbed experiments and simulation runs. We observe a good fit between experimental and simulation results.

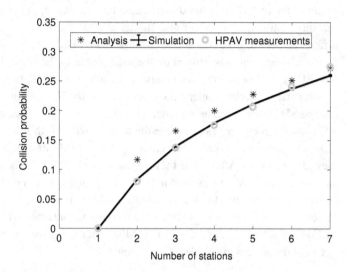

Figure 6.3 Collision probability obtained by simulation, model, and experiments with HomePlug AV devices for the default class CA1 of IEEE 1901 PLC standard given in Table 3.4.

Experimental Guideline To reproduce the testbed measurements above, we can run the following commands (see also Section 4.7.2):

```
1  ampstat -i <ETH_INTERFACE> -C -d <LINK_DIRECTION> -s <LINK_
   ID> -p <PEER_NODE_MAC_ADDRESS> # reset statistics at
   beginning of test
2  # start iperf server and clients in the appropriate devices
   , and measure throughput for a pre-determined duration
3  ampstat -i <ETH_INTERFACE>  -d <LINK_DIRECTION> -s <LINK_ID
   > -v -p <PEER_NODE_MAC_ADDRESS> # save statistics to file
   , with verbose mode that prints byte contents of the 0xA
   031 message
```

The ampstat command in line 3 is run in verbose mode and returns the bytes of the MM response. The number of acknowledged frames (included collided ones) is in bytes 25–32 and the number of collided frames is in bytes 33–40. We have to repeat the commands for all the stations transmitting traffic in the experiment.

6.3.2   Throughput and Overheads

We use the same testbed of eight stations, and we run experiments with the default configuration of IEEE 1901 standard (i.e., CA1). Both Wi-Fi and PLC MAC layers introduce some overhead, due to backoff delays, headers, preambles, management messages, acknowledgments, and interframe spaces. PLC devices are subject to an additional overhead, due to the presence of MMs (described in Chapter 4). These MMs are required for updating the modulation schemes (tone maps) between any two communicating stations. Both Wi-Fi and HomePlug AV use OFDM on the PHY layer. The difference between the two is that in Wi-Fi all OFDM subcarriers use the same modulation scheme, which is defined by the MCS index in the frame header [8], whereas in HomePlug AV, each carrier can employ a different scheme, because different frequencies experience variable attenuation. As a result, the HomePlug AV stations have to exchange MMs that indicate the modulation per carrier (917 used carriers for HomePlug AV), and the receiver has to update the transmitter with this information each time the channel conditions (bit error rate) change [6]. The MMs are sent with priority CA2/CA3, they also use CSMA/CA, and they consume backoff time. The exact amount of messages exchanged is not specified in the standard and depends on the implementation.

We run experiments to measure MAC and MMs overheads as a function of the number of stations N. Figure 6.4 shows the throughput obtained with Wi-Fi (802.11n) and PLC (HomePlug AV), and the PLC MM overheads (measured as the fraction of management frames among all data frames sent).

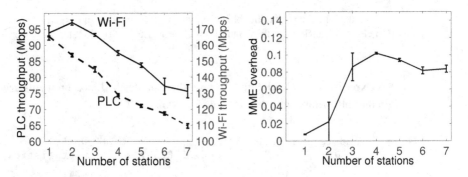

Figure 6.4 Total throughput of N saturated stations, $1 \leq N \leq 7$, from experiments with HomePlug AV PLC and 802.11n Wi-Fi devices (left). MMs overhead of the HomePlug AV tests (right). The measured PHY rate for PLC is ~150 Mbps and for Wi-Fi is ~300 Mbps. Both protocols include a large MAC overhead (about 40 percent), with HomePlug AV having the additional overhead of MMs, as N increases.

Both MACs follow relatively similar trends for throughput degradation when N grows. However, for $N \geq 3$, the PLC MMs are about 10 percent of all frames sent. Furthermore, for $N = 1$ station, we can compute the MAC layer overhead by computing the ratio of received throughput at the MAC layer and the physical data rates used on the links. We find similar overheads for Wi-Fi (about 40 percent) and PLC (about 38 percent). However, the MM measurements indicate that a significant fraction of the overhead in PLC can be due to MMs. As the channel conditions are ideal in our tests, we expect that this overhead can be larger for lossy links or links with electrical devices creating impulsive noise, which would need a faster tone map adaptation and more MMs.

📲 **Experimental Guideline**   To reproduce these testbed measurements, we can run the following commands (see also Section 4.7.4):

```
1  ampstat -i <ETH_INTERFACE> -C -d 0 -s <LINK_ID> -p <PEER_
   NODE_MAC_ADDRESS> # reset the statistics for all
   priorities (LINK_ID) 0--3 before the test
2  # start iperf server and clients in the appropriate devices
   , and measure throughput for a pre-determined duration
3  ampstat -i <ETH_INTERFACE> -d 0 -s <LINK_ID -v -p <PEER_
   NODE_MAC_ADDRESS> # save the output of this command to a
   file, to count the packets per priority LINK_ID
```

The ampstat command in line 3 measures the total number of MPDUs in the transmission direction ("-d 0" option). We have to repeat the above commands for all the stations transmitting traffic during the experiment. For each station, we have to repeat the commands for three LINK_ID values: data are run by defaut at CA1, hence LINK_ID is equal to 1. MMs use LINK_ID 2 and 3. 📲

6.3.3   PLC MAC Configuration

We evaluate the performance of different configurations for the PLC CSMA/CA not necessarily defined in the IEEE 1901 standard, using simulation and analysis. For the simulations, we use the network assumptions of Section 6.2.1 and the time slot duration and timing parameters specified by the IEEE 1901 standard. The data of a PLC frame transmission has a duration D, which is set to the maximum defined by IEEE 1901, since all stations are saturated, and is preceded by two priority resolution slots (PRS) and a preamble (P). It is followed by a response interframe space ($RIFS$), the ACK, and finally, the contention interframe space ($CIFS$) (see Section 3.5.1). Thus, a successful transmission has a duration $T_s = 2PRS + P + D + RIFS + ACK + CIFS$. In case of a collision, the stations defer their transmission for $EIFS$, where $EIFS$ is the extended interframe space defined by IEEE 1901; hence a collision has duration $T_c = EIFS$. Table 6.1 shows all the timing parameters we use to evaluate throughput given Equation (6.12).

Table 6.1 Simulation parameters

Parameter	Duration (μs)
Slot σ, priority slot PRS	35.84
$CIFS$	100.00
$RIFS$	140.00
Preamble P, ACK	110.48
Frame duration D	2500.00
$EIFS$	2920.64

Figure 6.5 compares the normalized throughput of PLC MAC obtained via simulation and model, for (1) the default parameters for the two priority classes CA1 and CA3 (CA0 and CA2 are equivalent according to Table 3.4), and (2) additional configurations.

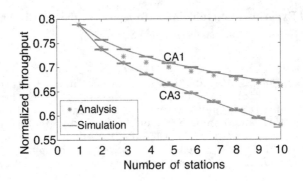

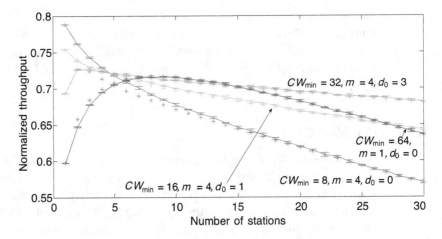

Figure 6.5 Normalized throughput obtained by simulation and model, for the default configurations of IEEE 1901 (top) and for various other configurations with $CW_i = 2^i CW_{min}$, and $d_i = 2^i(d_0 + 1) - 1$, $i \in \{0, m - 1\}$ (bottom). Lines represent simulations, points the analytical results.

There is a good fit between analysis and simulation, with an exception for small N and small d_is (CA0/CA1 class and $CW_{min} = 8$, $d_0 = 0$, $m = 4$ configuration). The reduced accuracy for small N in these cases is due to the decoupling assumption, which might fail to capture the coupling introduced by the deferral counter.

Figure 6.5 presents the trade-offs between parameters CW_{min}, m, d_0 and their effect on normalized throughput under different contention regimes. Recall that CA3 class uses smaller contention-window values than CA1 (see Table 3.4). Hence, under heavy contention, the probability that two or more stations select the same backoff counter value and collide is higher for CA3 than CA1. CA3 class should not be used when traffic demands are high, as it can cause delays due to collisions. As we discuss in Section 3.5, the parameters CW_{min}, m, d_0 affect the PLC multiuser performance and fairness. This is also evident in simulations of Figure 6.5. For instance, a $CW_{min} = 64$ value can lead to high delay access and low throughput when the number of stations is small, as they have to wait for $CW_{min}/2$ slots on average before accessing the channel. Similarly, a small number of backoff stages $m = 1$ can lead to low throughput as the number of users increases, because the stations do not double the contention window (hence do not react into lower transmission probability) after a collision.

6.4 PHY/MAC Cross-Layer Performance Evaluation

6.4.1 Rate Anomaly

In all previous sections, we have assumed that the PHY layer is ideal when stations contend. However, in practice, stations have different link qualities and PHY rates. Different rates can yield a performance anomaly, where the slowest station limits the total network throughput due to longer transmission duration compared to other stations. This phenomenon has been observed for early Wi-Fi standards [9]. Following our analysis in Equation 6.12, the throughput of any station in the network can be bounded by the duration of a collision or success of other stations.

We now examine whether rate anomaly exists for PLC, which would occur when the throughput of all contending links is bounded by the slowest transmission rate. We repeat experiments with links that differ in PHY rates by at least 50 percent (when they are run isolated). We use the testbed presented in Figure 5.1. Figure 6.6 presents throughput measurements of two and three links contending simultaneously, along with the corresponding metrics when the links are isolated. We observe that the throughput of the best link degrades at most 50 percent in the case of two contending links and 33 percent in the three links example, as expected in fair CSMA/CA protocol where stations share the airtime. Hence, rate anomaly does not seem to appear in PLC networks. The reason is that IEEE 1901 standard restricts the maximum frame in duration (μs) and not in size (bytes), using the frame aggregation presented in Section 3.3. As a result, the frame duration might be the same for saturated low-throughput and high-throughput links. The difference between any two links is the bit loading (BLE) and the bytes included in the frames. In addition to this result, we observe that the best link might dominate and gain higher throughput than half its available throughput (when isolated).

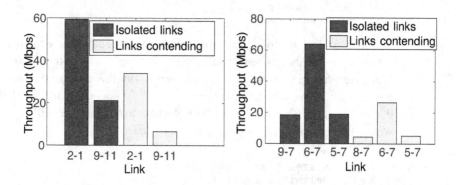

Figure 6.6 Performance of two contending links (left) and three contending links (right).

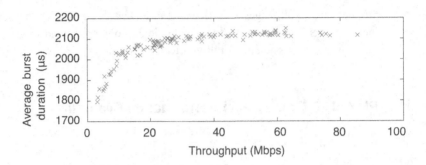

Figure 6.7 Burst duration versus throughput for all testbed links of Figure 5.1.

To justify the two observations above, we measure the average burst durations of all saturated links in the testbed. Figure 6.7 shows the average duration of the burst of frames versus the average throughput for every links. We observe that this average burst duration is almost independent of the link quality except from bad links. This can be one of the reasons why the best link dominates during contention of multiple flows: when a collision occurs, the link with the largest frame duration still manages to recover some of the PBs contrary to the other links that have all PBs collided. Also, since each symbol of the best link contains much more PBs than the bad links (the bit loading estimate for the best link is higher), even if the difference between their frame durations is a couple of symbols, the good link can recover tens of PBs. Another reason for the dominance of the best link can be a capture effect of the station with the highest SNR, which still can decode a few PBs even if these overlap with those of a link with lowest SNR.

💻 **Experimental Guideline**   To measure the average frame duration, we enable the sniffer mode in PLC devices and capture the frame control. We can use the following command and save the output to a file:

```
1       faifa -m -i <ETH_INTERFACE>
2       Choose the frame type (Ctrl-C to exit): 0xa034
3       Frame: Sniffer Mode Request (0xA034)
4       Sniffer mode?
5       0: disable
6       1: enable
7       2: no change
```

Select mode 1 above to enable sniffer mode. At the end of the experiment, you can re-run the command above to disable the sniffer mode with option 0, to avoid frame-capturing overheads on the device. It is best to run this command at the station that transmits or receives the data, as other stations might not be able receive all the MPDUs due to channel quality differences. Then, we have to compute the "HPAV frame length" with delimiter type 1, which represents data frames:

```
1       Direction: Rx
2       System time: 26657900886
3       Delimiter type: 1
4       STEI: 03
5       DTEI: 06
```

```
6      Link ID: 01
7      HPAV version 1.0: No
8      HPAV version 1.1: No
9      Bit loading estimate: 26
10     HPAV frame length: 530
11     MPDU count: 1
```

The frame length in line 10 above is in hexadecimal format and in multiples of 1.28 μs.

6.4.2   Hidden Terminals

The PHY layer of PLC is constantly advancing toward high data rates. Enterprise and residential buildings are expected to host different PLC networks with many stations, hence multiple collision domains. A remaining important challenge for the scalability of PLC networks is to achieve efficient resource allocation in multiple collision domains, as performance degrades due to hidden terminals which arise when stations cannot listen to each other's transmissions but still their packet transmissions collide at their receivers (see Section 3.6.2 for definition). The IEEE 1901 standard specifies that the RTS/CTS mechanism or an optional TDMA mechanism can be used to overcome hidden node scenarios [6]. However, by capturing the SoF (see experimental guidelines above: we did not capture any delimitery type equal to 3, which is the RTS/CTS), we find that RTS/CTS nor TDMA are not used in commercial chipsets. Note that in PLC, all stations utilize the whole available bandwidth compared to Wi-Fi standards that include multiple channels with different central frequencies [8, 10].

A hidden terminal problem can arise even on a building of the dimensions of our testbed in Figure 5.1. Figure 6.8 shows two representative links contending with UDP and TCP traffic. As we observe, the throughput of contending links is much lower than half of their throughput in isolation, which would be expected if the stations could hear each other. We observe that situations with hidden nodes arise in our testbed and that the performance severely degrades in these cases. For the next-generation PLC technologies to scale, RTS/CTS mechanism and introduction of channelization are necessary for reducing the number of hidden terminals.

6.5   Short-Term Fairness

In the previous sections, we discuss the throughput figures of the PLC CSMA/CA protocol, focusing on average performance and not on short timescales. We now evaluate the short-term dynamics of PLC multiuser performance, considering delay-sensitive applications that depend on this short-term dynamics. We will see that a high throughput of PLC CSMA/CA – introduced by the deferral counter – can come at a cost of short-term unfairness and high delay variance.

When the medium or transmission opportunities are shared equally among the stations during a fixed time interval, the MAC protocol is considered to be fair. Depending

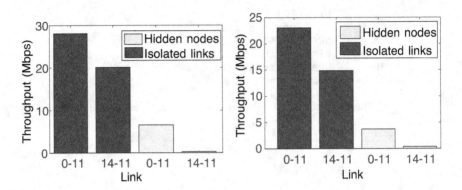

Figure 6.8 Performance of two contending links with hidden terminals transmitting UDP (left) and TCP (right) traffic.

on the time horizon, fairness is defined as long term or as short term, and it affects the quality of service of the application running on the device. Long-term fairness benefits the average throughput per user, but short-term fairness can affect delay variance (i.e., jitter). In a short-term unfair system, a station might have to wait for many other stations' transmissions before it transmits again, which results in high delay variance. Short-term fairness is a stronger system property than long-term fairness, because short-term fairness implies long-term fairness, but the reverse does not necessarily hold.

TDMA can be optimal protocol to share the medium in terms of short-term fairness, as the PLC CCo can distribute equally airtime to users. IEEE 1901 supports TDMA (see Sections 3.5 and 3.6.1), but the default access method used in practice is CSMA/CA, as we observe from commercial devices. TDMA might not practical under highly dynamic traffic, as it requires a centralized authority to allocate time per user. For time allocation, is also needs MM transmissions containing the data demands per station, which introduce overhead.

We first examine snapshot of the short-term dynamics of Wi-Fi and PLC from testbed experiments. Figure 6.9 shows the potential unfairness and jitter phenomena of PLC, in a testbed with two stations. Compared to Wi-Fi, PLC can be short-term unfair. While PLC Station A transmits during several consecutive slots, Station B is likely to remain in a state where it has a high probability of colliding or sensing the medium busy. Station B is then even less likely to transmit while in this state, and it might have to wait several tens of milliseconds before the situation reverts. This unfairness causes a high delay-variance (i.e., high jitter), because a station can transmit many times consecutively with a low latency and then can wait for a long time with high latency.

The fairness issue with PLC is due to the introduction of the deferral counter, because it causes stations to double the contention window after sensing the medium busy. In networks of a few stations, the last station that transmitted has a higher probability of transmitting again compared to other stations, because it has a smaller contention window. Any station that transmits, moves to backoff stage 0 and sets the contention window to the minimum one. However, the rest of the stations might double their contention

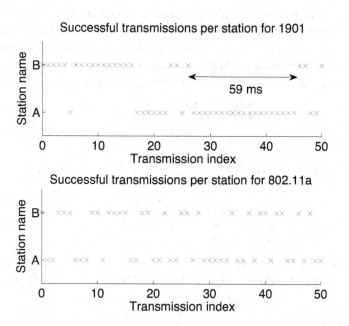

Figure 6.9 Trace of fifty successful transmissions by two saturated stations with PLC (1901) and Wi-Fi (802.11a mode). PLC exhibits short-term unfairness: a station holding the channel is likely to continue holding it for many consecutive transmissions, which causes high jitter.

window due to the deferral counter. As we observed in Section 6.3, the deferral counter can adjust the normalized throughput by configuring its initial values at each backoff stage. The deferral counter can also provide different fairness levels depending on its initial values. We later discuss how the deferral counter creates a trade-off between throughput and fairness.

Experimental Guideline In the test above, we set both stations to transmit `iperf` UDP traffic to the same third station. To repeat the above experiment, we enable the sniffer mode in the PLC receiver and capture the MPDU frame control. For Wi-Fi, we can capture the Ethernet frames at the receiver using `tcpdump` utility; 802.11a does not use frame aggregation, hence every Ethernet frame corresponds to a single Wi-Fi transmission. For PLC, we cannot use `tcpdump`, because frame aggregation yields a variable number of Ethernet packets per PLC frame. We can use the following command at the receiver of the traffic and save the output to a file:

```
1    faifa -m -i <ETH_INTERFACE>
2    Choose the frame type (Ctrl-C to exit): 0xa034
3    Frame: Sniffer Mode Request (0xA034)
4    Sniffer mode?
5    0: disable
6    1: enable
7    2: no change
```

Select mode 1 above to enable sniffer mode. At the end of the experiment, you can re-run the command above to disable the sniffer mode with option 0, in order to avoid frame capturing overheads on the device. It is best to run this command at the station that receives the data, as other stations might not be able receive all the MPDUs due to channel quality differences. Then, we have to save the sequence of STEI value of all MPDUs (line 4 below), which is the identification of the data source station.

```
1         Direction: Rx
2         System time: 26657900886
3         Delimiter type: 1
4         STEI: 03
5         DTEI: 06
6         Link ID: 01
7         HPAV version 1.0: No
8         HPAV version 1.1: No
9         Bit loading estimate: 26
10        HPAV frame length: 530
11        MPDU count: 1
```

During the analysis of this trace, it is important to remember that stations contend for bursts of MPDUs, hence consecutive MPDUs with MPDU count 0–2 (line 11 above) should be considered as a single burst and CSMA/CA transmission.

6.5.1   Metrics of Fairness

We analyze and evaluate fairness of the PLC protocol with two metrics: Jain's fairness index and the number of intertransmissions. Both metrics have been also used in the fairness analysis of Wi-Fi (e.g., [11, 12]).

Jain's fairness index represents the variability of a set of measurements, which here is the throughput of PLC stations [13]. The larger the values of Jain's index J are, the better is the fairness. Ideally, when all stations share equally the channel, Jain's index is equal to 1. In MAC layer performance evaluation, we can define it as follows.

DEFINITION 6.1. *Let N be the number of contending stations in the network and ρ_n be the throughput of station n within a time period. Then, Jain's fairness index is defined as*

$$J = \frac{\left(\sum_{n=1}^{N} \rho_n\right)^2}{N \sum_{n=1}^{N} \rho_n^2}, \quad 1/N \le J \le 1. \tag{6.13}$$

Fairness can be evaluated at different time horizons. As the time horizon increases, fairness should improve in a CSMA/CA protocol with homogeneous users, since by the law of large numbers, every user gets an equal share of the channel in the long run. To analyze fairness as a function of the time horizon, we can use the sliding-window method (SWM) introduced by Koksal et al. [14]. SWM slides a window of w packets across a trace of successful transmissions. For each w packets, Jain's fairness index can be computed from (6.13), with ρ_n being the fraction of transmissions performed by station n. Jain's fairness index $J(w)$ with window size w is the average of Jain's index

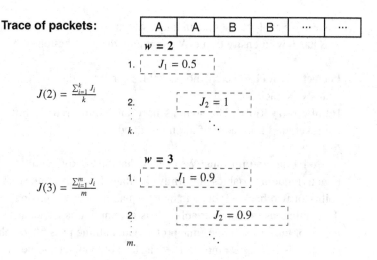

Figure 6.10 Sliding-window method example with two contending stations A and B. The label on a successfully transmitted packet represents the source of the packet. $J(w)$ is computed for two window values $w = 2$ and $w = 3$.

values for each consecutive sequence of w packets in the trace. The fairness time horizon increases with w. Figure 6.10 shows an example of the SWM for two stations and two w values. We can use a normalized window W with respect to number of stations N, given by $W = w/N$. Jain's fairness index with the SWM is a metric of fairness on different timescales. However, it is analytically intractable. Therefore, to model short-term fairness, we can use another metric, called the number of intertransmissions.

The number of intertransmissions was introduced by Berger-Sabbatel et al. [11] for the short-term fairness of the Wi-Fi CSMA/CA.

DEFINITION 6.2. *The number of intertransmissions K for a tagged station B is the number of transmissions by other stations between two consecutive transmissions of B [11].*

According to the definition above, if $K = 0$, then after a successful transmission of Station B, the next transmission is done again by B. If $K = 1$, Station B transmits once, then another station transmits, and then B transmits again. If there are N stations contending for the channel, the ideal value of K for short-term fairness is $N - 1$. In practice, K is a random variable, and $\mathbb{E}[K] = N - 1$ represents an ideally fair protocol. The second moment of K is related to jitter: the larger $\mathrm{Var}(K)$, the higher are the delay variance and jitter.

6.5.2   Mathematical Analysis: The Distribution of Intertransmissions

We use similar networking assumptions as for modeling throughput:

Perfect sensing: There are two stations A and B that can always receive each other's transmissions, and there are no hidden terminals.

Initial system state: Both stations start at backoff stage 0. From a practical point of view, we analyze the system under scenarios where Station A is saturated and Station B has low-intensity traffic. We compute the distribution of K for the tagged Station B.

Perfect channel: There is no packet loss or errors due to PHY layer, which makes the analysis tractable.

Infinite retry limit: The stations have an infinite retry limit: They never discard a packet until it is successfully transmitted.

Modeling assumptions: We assume that the backoff counters of the stations are continuous random variables uniformly distributed over the range $[0, CW]$. Hence, the probability of two backoff counters being equal, thus of a collision, is equal to 0 thanks to the continuous random variables. This assumption is necessary because the analytical model of the real discrete-time protocol and all the possible combinations of transmissions/collisions appear intractable. The backoff counters of the two stations are independent, which follows the real protocol where stations select independently their backoff counters. The computation of the distribution $\mathbb{P}(K = k)$ is based on the fact that, whenever the stations choose or resume their backoff counters, the station with the smallest backoff counter transmits.

Let A and B be two contending stations. We tag station B. Let b be the backoff counter of B and a_i be the backoff counter of A before its ith transmission. Then, the distribution of K for Wi-Fi is given by [11]

$$\mathbb{P}(K = k) = \mathbb{P}\left(\sum_{i=1}^{k} a_i < b \text{ and } \sum_{i=1}^{k+1} a_i \geq b\right), \text{ or } \mathbb{P}_{802.11}(K = k) = \frac{k+1}{(k+2)!}, \ k \geq 0. \quad (6.14)$$

The computation of $\mathbb{P}(K = k)$ for PLC is more complex than it is for Wi-Fi. Let BC_B, CW_B, and DC_B be the values of backoff counter, contention window, and deferral counter at Station B, respectively. We define similarly the counters of Station A. With Wi-Fi, under the collision-free assumption, the contention window remains constant for both stations, independently of the value of k, whereas in PLC, CW_B is doubled whenever A transmits and $DC_B = 0$, because B senses the medium busy. To evaluate $\mathbb{P}_{1901}(K = k)$ for the default priority class CA1 of IEEE 1901, we compute the probability that A transmits k times consecutively, or equivalently the probability that $BC_B < BC_A$ at each of the k transmission attempts. BC_A is always uniformly distributed over the range $[0, 8]$, whereas BC_B depends on k. When $k = 0$, we have $CW_B = 8$ and $DC_B = 0$. After each transmission of A, BC_B is updated as follows: if DC_B is equal to 0, then new BC_B and DC_B are chosen depending on the backoff stage of station B; otherwise, BC_B is decremented by BC_A (Station B has already counted down BC_A idle slots) and DC_B is decremented by 1. By using this reasoning and Table 3.4 with the 1901 parameters, we compute DC_B and the range to which BC_B belongs for each k. The distribution of K for the CA1 priority class of IEEE 1901 is given as follows, and the computations can be found in Appendix B.

PROPOSITION 6.1. *The distribution of the number of intertransmissions K for the CA1* *priority class of IEEE 1901 standard is*

$$
\mathbb{P}_{1901}(K = k) = \begin{cases}
1/2, & k = 0, \\
1/8, & k = 1, 2, \\
1/32, & 3 \le k \le 6, \\
0.0578^l/128, & 7 \le k - 16l \le 14, \\
0.0578^l \cdot \mathbf{I}(k - 16l - 14)/8, & 15 \le k - 16l \le 22,
\end{cases}
\tag{6.15}
$$

where $l = \lfloor (k - 7)/16 \rfloor$, and $\mathbf{I}(n)$ is the nth entry (for n between 1 and 8) of the vector $\mathbf{I} = 10^{-2}(6.25\ 6.25\ 6.24\ 6.17\ 5.92\ 5.38\ 4.53\ 3.49)$ [15].

Comparing the plots of Proposition 6.1 and Equation 6.14 in Figure 6.11, we see that the distribution of K has a heavier tail for PLC than for Wi-Fi. This means that a PLC station is more likely to perform many more consecutive back-to-back transmissions than a Wi-Fi one. To show this mathematically, consider the survival functions of the 1901 (PLC) and 802.11 (Wi-Fi) distributions of K, $S_{1901}(k) = \mathbb{P}_{1901}(K > k)$ and $S_{802.11}(k) = \mathbb{P}_{802.11}(K > k)$. From (6.14) and (6.15), we have

$$
\lim_{k \to \infty} \frac{S_{1901}(k)}{S_{802.11}(k)} = \infty,
\tag{6.16}
$$

which shows that the tail for PLC is heavier than for Wi-Fi [16]. This is also observed from the first and second moments of the distributions: we have $\mathbb{E}[K] = 2.8$ and $\mathrm{Var}(K) = 28.28$ for PLC, whereas Wi-Fi yields $\mathbb{E}[K] = 0.73$ and $\mathrm{Var}(K) = 0.77$. The variance explodes with PLC. Therefore the IEEE 1901 standard default CA1 configuration generates much higher jitter than Wi-Fi.

The distribution of K is similar for any minimum contention-window value CW_0, as long as the deferral counter has the same values and the contention window is doubled between successive backoff stages. To study the realistic PLC protocol with discrete backoff counters, we run simulations with increasing values of CW_0, multiplying CW_0 by 2 and maintaining the deferral-counter values the same as in Table 3.4. For $CW_0 = 8$ (CA1 class of IEEE 1901), we find that $\mathbb{E}[K] = 4.2$ and $\mathrm{Var}(K) = 70.1$. Figure 6.12 depicts $\mathbb{P}_{1901}(K = k)$ from analysis and simulations of the real protocol with $CW_0 = 8$. We observe that there is some deviation in $\mathbb{P}_{1901}(K = 2)$, which translates into a deviation of $\mathbb{E}[K]$ between analysis and simulation, due to the long-tailed distribution. As $CW_0 \to \infty$, the collision probability tends to 0, and the discrete distribution of the backoff counters (real PLC protocol) can be approximated by the continuous one (model of intertransmissions).

Extension of the model to $N > 2$: We have modelled a network with $N = 2$ PLC stations, because the simplicity of the transmission sequences enables us to maintain the uniform distribution of the backoff counters, following the IEEE 1901 standard. Hence we can achieve a high level of accuracy with the analysis. For $N > 2$, due to the challenging nature of the modeling problem, we need to assume exponential backoff counters that simplify the computation of the minimum backoff counter that wins the transmission. In such case, if the mean of the exponential backoff counter at station i is $1/\lambda_i$, then a station i wins the channel with probability $\mathbb{P}\left(i | BC_i = \min_{n=1}^{N} BC_n\right) = \lambda_i / \sum_{n=1}^{N} \lambda_n$.

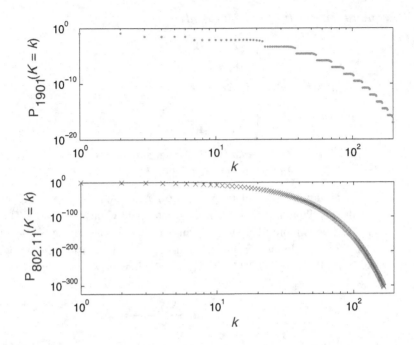

Figure 6.11 The distribution of K for IEEE 1901 (CA1 priority) and IEEE 802.11 computed analytically in log-log scale. The x-axis is the same for both plots. Note the difference in y-axis scale between the two standards, e.g., $\mathbb{P}_{1901}(K = 15) \sim 10^{-3}$ and $\mathbb{P}_{802.11}(K = 15) \sim 10^{-14}$. The IEEE 1901 distribution of K attenuates periodically (see Equation (6.15)).

The distribution of the minimum of N exponential random variables is another exponential random variable with mean $1/\sum_{n=1}^{N} \lambda_n$. Even with this exponential approximation, the state space for $N > 2$ explodes: the first terms of the probability mass function of the distribution $\mathbb{P}(K = k)$ can be computed, but the model is intractable for large k due to the explosion of the number of possible states (sequences of transmissions) in the time horizon. The distribution can be computed numerically by using combinatorics and grouping sequences with the same probability distributions to reduce the run-time.

6.5.3   Jain's Fairness Index Evaluation

We use simulations and experiments with commercial chipsets to compare the short-term and long-term fairness of PLC and Wi-Fi. To compare the protocols, we use a `Matlab` simulator similar to the one of Section 6.3 to obtain a trace of successfully transmitted frames over the channel. We do not simulate the PHY layer, because the two channels are different and to avoid hidden factors that might affect a CSMA/CA evaluation. In addition to the simulator, we have testbed of six stations with HomePlug AV and 802.11a interfaces under ideal channel conditions. In this setup, up to $N = 5$ stations send saturated UDP traffic to the $(N + 1)$th station using `iperf`. The $(N + 1)$th

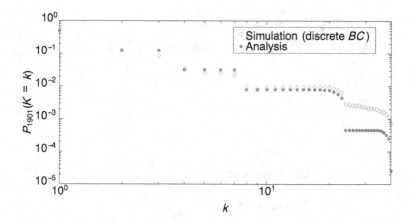

Figure 6.12 The distribution of K for 1901 (CA1 priority) in log-log scale computed analytically and via simulation with the real discrete backoff counters BC.

station captures the frames received either using `faifa` or `tcpdump` (see experimental guideline in Section 6.5).

Figure 6.13 shows the results of the SWM applied to traces of successfully transmitted packets from simulation and experiments for $N = 2$. Jain's fairness index $J(W)$ for 802.11a is higher than for 1901 for all values of W. For example, $J(W) = 0.95$ is achieved with $W = 5$ for 802.11a and with $W = 70$ for 1901. Thus Wi-Fi CSMA/CA is fairer than PLC for all time horizons, and particularly in the short term.

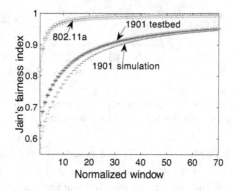

Figure 6.13 Jain's fairness index with the SWM. Simulations coincide well with testbed results. For 802.11a, simulation and testbed curves are indistinguishable.

The short-term fairness of the protocols can change as N increases. Figures 6.14a and 6.14b show $J(W)$ for $N = 3$ and 5, respectively. The two curves for 1901 and 802.11a approach each other as N increases, if we compare Figures 6.13, 6.14a, and 6.14b. The largest difference between the two CSMA/CA protocols appears for $N = 2$. As N increases, there is a higher probability that some station will steal the medium from the station already monopolizing it in PLC.

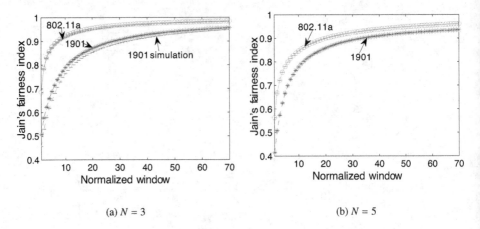

(a) $N = 3$ (b) $N = 5$

Figure 6.14 Testbed results of Jain's fairness index with the SWM for $N = 3, 5$.

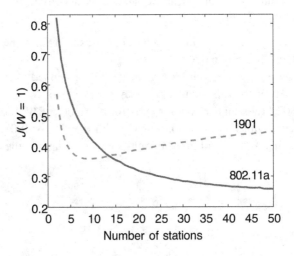

Figure 6.15 $J(W = 1)$ as a function of N from simulation results. Short-term fairness improves for 1901 and deteriorates for 802.11 as N increases.

Figure 6.15 plots $J(W = 1)$ as a function of N in simulation. $J(W = 1)$ for 802.11a decreases as N increases, which has already been observed in previous works [11, 12]. This is due to the collision probability that increases with N. In 802.11 for large N, after a collision, a station that transmitted successfully in the past is favored compared to the others, because there exists a subset of stations with a larger CW due to collisions. Figure 6.15 also illustrates that 1901 can be fairer than 802.11 in the short term for $N \geq 15$. Furthermore, it shows two regimes that can characterize the 1901 behavior; for $N < 10$, $J(1)$ decreases, and for $N \geq 10$, $J(1)$ increases. When $N < 10$, $J(1)$ for 1901 decreases for the same reason as for 802.11, as explained above. When $N \geq 10$, $J(1)$ increases, because under high-collision-probability conditions, in 1901 there might not exist a subset of stations with low CW, which is contrary to 802.11: after a collision, all stations (including those that transmitted successfully in the past) increase their CW

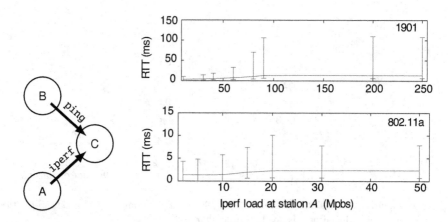

Figure 6.16 Illustration of the testbed experiment (left). The median RTT of 10^4 ping requests of B as a function of A's `iperf` load (right). The error bars represent the 10th and 99th percentiles. Observe the difference in the scales of y-axes of 802.11 and 1901.

due to sensing the medium busy. Moreover, the increasing behavior of $J(1)$ for 1901, when N is large, is attributed to the high collision-probability and to the inability of the maximum contention window $CW_{max} = 64$ to accommodate collisions. Therefore, when N is large, the majority of the stations have $CW = CW_{max}$, hence they have the same probability of transmitting, which results in fairness but a higher collision rate with respect to Wi-Fi that has a $CW_{max} = 1023$ value.

6.5.4   Short-Term Fairness and Impact on Delay

We discuss why short-term fairness is important in any multiuser setting with a realistic example of the two stations test presented in Figure 6.16. Station A sends UDP traffic with a varying load to a station C by using `iperf`, while Station B sends ping requests of size 1400 bytes to C. This represents a scenario with a bandwidth-hungry application (Station A) and a low-demand and delay-sensitive one (Station B) contending simultaneously for the medium. Figure 6.16 shows the median round trip time (RTT) of 10 000 ping requests of Station B for both 1901 (top) and 802.11a (bottom). The maximum RTT was 215.7 ms for 1901 and 11.5 ms for 802.11 when Station A was saturated. The results show that in a scenario where Station A sends a high data load and Station B sends smaller-size bursty traffic, Station B experiences larger delays than the maximum tolerable delays of delay-sensitive applications. The average and maximum RTT of Station B when Station A is saturated measured 8.2 and 21 ms, respectively, for CA3 priority.

We compare the two priority classes of the 1901 standard (see Table 3.4) in simulation when all stations are saturated in Figure 6.17. We find that the delay-sensitive class of the standard CA3 yields better fairness and lower jitter than CA1. In Figure 6.5, we saw that the normalized throughput for the CA3 class is lower than that of CA1. Therefore there is a trade-off that is already exploited in the IEEE 1901 standard for different levels of quality of service (QoS).

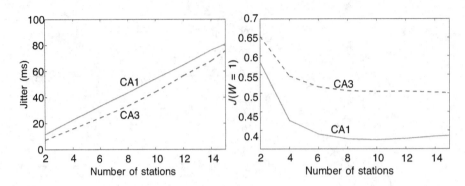

Figure 6.17 Simulations with the IEEE 1901 configurations for CA1 and CA3 priority classes. The better the fairness is, the lower the jitter is.

6.6   The Trade-off between Throughput and Short-Term Fairness

We explore the trade-off between throughput and fairness in PLC CSMA/CA for a range of configurations with $CW_i = 2^i CW_{min}$, and $d_i = f^i(d_0 + 1) - 1$, $i \in \{0, m - 1\}$. To measure short-term fairness, we compute Jain's fairness index over windows of N packet durations, and we use $J(W = 1)$ (see Section 6.5.1 for details). We take the normalized $W = 1$, as this is the smallest value of W such that $J(W)$ can be equal to 1 (for a perfectly fair protocol).

Figure 6.18 plots normalized throughput and short-term fairness as a function of the initial values of the deferral counter (in terms of d_0 and f), for $N = 2$ and $N = 5$. A direct link can be made between throughput and fairness. When the initial deferral-counter values are small (small values of d_0 and f), the stations are more likely to react upon sensing the medium busy and thus proactively avoid collisions and airtime wastage. Conversely, these configurations have bad short-term fairness and thus can cause high jitter.

We look in more detail at the throughput/fairness trade-off in Figures 6.19 and 6.20. Figure 6.19 shows throughput and fairness for various initial deferral counter values (in terms of d_0 and f values). Figure 6.20 considers these two metrics for different numbers of backoff stages m. Both figures show a trade-off between throughput and short-term fairness. This trade-off can be tuned by adapting the number of backoff stages and the initial values of the deferral counter. This is a feature of PLC MAC, enabled by the deferral counter. We summarize the effects of all parameters on throughput and fairness in Table 6.2, which also contains the minimum contention window CW_{min}. In general, as discussed in Section 6.5.2, CW_{min} does not affect fairness as long as the contention windows are doubled between successive backoff stages. However, CW_{min} can affect throughput: for small N, smaller CW_{min} values are better because of smaller backoff overhead; for large N, larger CW_{min} values are better for lower collision probability and higher throughput.

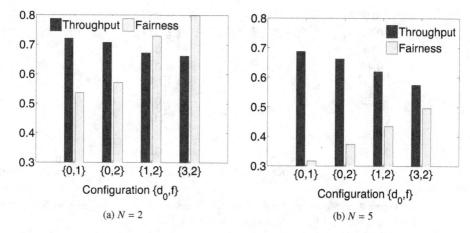

Figure 6.18 Throughput and fairness ($J(W = 1)$) with $CW_{min} = 8$ and $m = 4$ and various values of d_0 and f. The initial value of the deferral counter at backoff stage i is given by $d_i = f^i(d_0 + 1) - 1$.

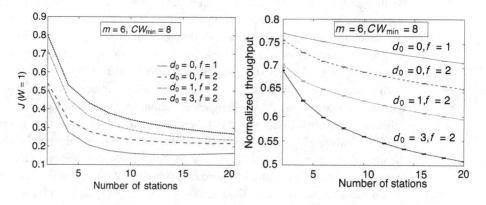

Figure 6.19 Short-term fairness and throughput obtained by simulation for parameters $CW_{min} = 8$, $m = 6$ and various values of d_0 and f. The deferral counter determines a trade-off between throughput and fairness in 1901.

Table 6.2 Summary of the qualitative effects of each PLC CSMA/CA parameter on throughput ("T") and short-term fairness ("F")

	d_0	f	m	CW_{min}
Small	T ↗ F ↘	T ↗ F ↘	T ↘ F ↗	T ↗ if N is small F →
Large	T ↘ F ↗	T ↘ F ↗	T ↗ F ↘	T ↗ if N is large F →

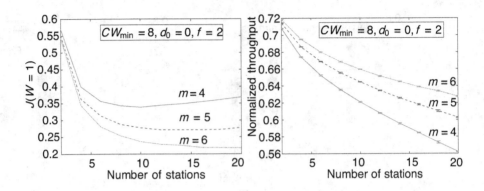

Figure 6.20 Short-term fairness and throughput obtained by simulation with parameters $CW_{min} = 8, d_0 = 0, f = 2$, and various values of m.

6.7   Fairness and Coupling between Stations

We discuss how short-term unfairness can reduce the accuracy of modeling hypotheses that assume independence between the backoff processes of the stations. Bianchi [1] introduces such an approximation to study the Wi-Fi MAC. His approximation works well for any number of stations. However, in Section 6.3, we observe that a similar assumption for PLC might yield inaccuracies for a small number of stations.

The decoupling assumption uses the approximation that the backoff processes of the stations are independent. As a consequence, stations experience the same time-invariant collision probability, independently of their own state and of the state of the other stations. To analyze PLC, it has been assumed that a station senses the medium busy with the same time-invariant probability (equal to the collision probability) at any time slot. However, the deferral counter can introduce some coupling among the stations: after a station gains access to the medium, it can keep it for many consecutive transmissions before any other station can transmit. As a result, the collision and busy probabilities might not be time-invariant for PLC networks.

Let X_i be the variable that indicates which station transmits successfully at the ith transmission for two stations A and B. If A transmits, we take $X_i := 1$, and if B transmits, we have $X_i := 2$. Figure 6.21 shows evidence of the 1901 unfairness from a testbed trace of 50 000 frame transmissions by using the autocorrelation function of the identification of station that transmits in a network of two saturated stations A and B. We show the autocorrelation function of X_i, $1 \leq i \leq 5 \cdot 10^4$. The autocorrelation is positive for 1901 at lags smaller than 15, which means that if $X_i = 1$ for some i, it is likely that $X_{i+1} = 1$. In contrast, for 802.11a we have a negative value of autocorrelation at lag 1 and a positive one at lag 2, which means that if $X_i = 1$ for some i, it is very likely that $X_{i+1} = 2$ and $X_{i+2} = 1$. The dependence of the nontransmitting station to the one monopolizing the medium can result in different collision probabilities per backoff stage. To see this, we plot the collision probabilities experienced by 802.11 and 1901 stations, as a function of the backoff stage (i.e., as a function of the stations' state) in Figure 6.22.

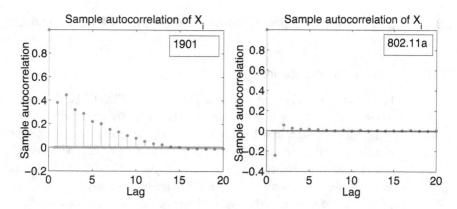

Figure 6.21 We study a testbed trace of successfully transmitted packets for both IEEE 1901 and IEEE 802.11a, when two saturated stations A and B contend for the medium. The autocorrelation function of the identification of station that transmits shows that 1901 is less fair than 802.11.

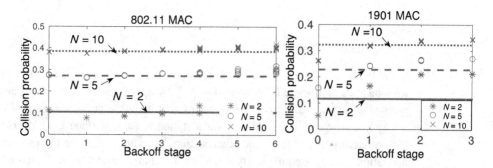

Figure 6.22 Simulations of 1901 (with CA1 parameters) and 802.11 for $N = 2, 5, 10$. Points show the collision probabilities at different backoff stages for all stations, and lines represent the solution of the fixed-point equations for the collision probability from decoupling assumption models. The decoupling assumption is valid for 802.11, even for $N = 2$, whereas the collision probability might depend on the backoff stage for 1901.

In the same figure, we also show the collision probabilities computed with the model introduced in Section 6.3. Let O_k be the sequence of outcomes of attempted transmissions, that is, $O_k \doteq 1$, if the kth transmission attempt results in a success, and $O_k \doteq 0$ when the outcome is a collision. The decoupling assumption asserts that the sequence $\{O_k\}$ consists of i.i.d. random variables. In Figure 6.22, we observe that for 1901, $\{O_k\}$ cannot be considered as i.i.d., because the collision probability observed at different backoff stages is not the same. Thus the collision probability depends on the previous transmission attempts or on other stations' activity (both of which have an effect on the station's current backoff stage). Figure 6.22 shows that the collision probability for 1901 increases with the backoff stage i. To address this inaccuracy, we can use another model that does not rely on the "time-invariant collision-probability" assumption.

6.8   ⛩ Further Reading: Coupled Analysis of PLC CSMA/CA

6.8.1   Modeling Assumptions

In this section, we present a coupled model for the PLC CSMA/CA protocol (without relying on the decoupling assumption). The analysis relies on the same networking assumptions as Section 6.2.1, that is, perfect sensing and channel, N saturated stations in a single collision domain, infinite retry limit, and a homogeneous network.

The network is modelled as a dynamical system that describes the expected change in the number of stations at each backoff stage between any two consecutive time slots. In the stationary regime, the expected number of stations at each backoff stage is constant, hence we can compute performance by finding the equilibrium of the dynamical system. Compared to the model in Section 6.2, where the collision probability γ is time invariant, here the collision probability depends on the station's state, using the findings of Figure 6.22. Hence, we now use the notation γ_i for the collision probability at backoff stage i.

6.8.2   Mathematical Analysis

Let m be the number of backoff stages, and let $n_i, 0 \leq i \leq m-1$ denote the number of stations at backoff stage i. We should have $\sum_{i=0}^{m-1} n_i = N$ and $n_i \in \mathbb{N}$. The transmission probability at stage i, τ_i, is the probability that a station at backoff stage i transmits at any given time slot. In addition, for a station at backoff stage i, we denote with γ_i the probability that at least one other station transmits. Let p_e be the probability that no station transmits equivalently – that the medium is idle. Under the assumption of independence of the transmission attempts, we have

$$p_e = \prod_{k=0}^{m-1} (1 - \tau_k)^{n_k}, \tag{6.17}$$

and

$$\gamma_i = 1 - \frac{p_e}{1 - \tau_i} = 1 - \frac{1}{1 - \tau_i} \prod_{k=0}^{m-1} (1 - \tau_k)^{n_k}. \tag{6.18}$$

Station model: For a station at backoff stage i, we assume that the event that some other station transmits in a slot occurs with a constant probability γ_i, independent of the station's backoff and deferral counters values. Hence, this is the probability that a transmission of the given station collides, as well as the probability that the station senses a slot busy when it does not transmit. With this model, we derive the probability that a station transmits and that it moves to the stage $i + 1$ due to the deferral counter at any random time slot.

By following a similar process as in Section 6.2.2, we can compute x_k^i, the probability that a station at backoff stage i jumps to the next stage $i + 1$ in k or fewer time slots due to sensing the medium busy, directly from γ_i. The only difference with the model

Table 6.3 Notation list relevant to a station at backoff stage i

Notation	Definition (at backoff stage i, $0 \le i \le m - 1$)
n_i	Number of stations
γ_i	Probability that at least one other station transmits at any slot
p_e	Probability that the medium is idle at any slot (independent of i)
x_k^i	Probability that a station leaves stage i due to sensing the medium busy $d_i + 1$ times during k slots
bc_i	Expected number of backoff slots
τ_i	Probability that a station transmits at any slot
β_i	Probability that, at any slot, a station leaves stage i due to sensing the medium busy $d_i + 1$ times
F_i	Expected change in n_i between two consecutive slots
$\bar{n}_i$	Expected number of stations

introduced in Section 6.2.2 is that now, the collision probability depends on the station's backoff stage i. This yields

$$x_k^i = \sum_{j=d_i+1}^{k} \binom{k}{j} \gamma_i^j (1 - \gamma_i)^{k-j}. \tag{6.19}$$

Similarly to Section 6.2.2, we use bc_i for the expected number of time slots spent by a station at backoff stage i, which now is a function of γ_i:

$$bc_i = \frac{1}{CW_i} \sum_{k=d_i+1}^{CW_i-1} \left[(k + 1)(1 - x_k^i) + \sum_{j=d_i+1}^{k} j(x_j^i - x_{j-1}^i) \right] + \frac{(d_i + 1)(d_i + 2)}{2CW_i}. \tag{6.20}$$

Transmission probability τ_i can be expressed as a function of x_k^i and bc_i, using the renewal-reward theorem, with the number of backoff slots spent in stage i being the renewal sequence and the number of transmission attempts (i.e., 0 or 1) being the reward (see also Section 6.2). The expected number of transmission attempts at stage i can be computed similarly to bc_i. By dividing the expected number of transmission attempts at stage i with the expected time slots spent at stage i, τ_i is given by

$$\tau_i = \frac{\sum_{k=d_i+1}^{CW_i-1} \frac{1}{CW_i}(1 - x_k^i) + \frac{d_i+1}{CW_i}}{bc_i}. \tag{6.21}$$

Similarly, we define β_i as the probability that, at any given slot, a station at stage i moves to the next backoff stage because it has sensed the medium busy $d_i + 1$ times:

$$\beta_i = \frac{\sum_{k=d_i+1}^{CW_i-1} \frac{1}{CW_i} \sum_{j=d_i+1}^{k} (x_j^i - x_{j-1}^i)}{bc_i}. \tag{6.22}$$

The notation is summarized in Table 6.3. We next study the evolution of the expected change (drift) in the number of stations at each backoff stage i.

6.8.2.1 Transient Analysis of the System

To study the system, we use a vector for the number of stations at each backoff stage. Let $\mathbf{X}(t) = (X_0(t), X_1(t), \ldots, X_{m-1}(t))$ represent the number of stations at each backoff stage $(0, 1, \ldots, m-1)$ at time slot t. We use the notation $\mathbf{n}(t) = (n_0(t), n_1(t), \ldots, n_{m-1}(t))$ to denote a realization of $\mathbf{X}(t)$ at some time slot t.

Network model: We rely on the simplifying assumption that a station transmits, or moves to the next backoff stage upon expiring the deferral counter, with a constant probability (independently of previous time slots). This is necessary as otherwise, we would need to keep track of the backoff and deferral counter values of each station and the model would become intractable. The assumptions are as follows: (1) a station at backoff stage i attempts a transmission in each time slot with a constant probability $\tau_i(\gamma_i)$; and (2) a station at backoff stage i moves to backoff stage $i+1$ due to the deferral counter expiration with a constant probability $\beta_i(\gamma_i)$ in each time slot where it does not transmit. Both τ_i and β_i depend on the probability γ_i that the station senses a slot busy, which is computed from the transmission probabilities of the other stations following (6.18). With the above assumptions, $\mathbf{X}(t)$ is a Markov chain. The transition probabilities τ_i and β_i depend on the state vector $\mathbf{n}(t)$ and they can be computed from (6.18), (6.21), and (6.22). To simplify notation, we drop the input variable t from $\gamma_i(t)$, $\tau_i(t)$, $\beta_i(t)$, and $\mathbf{n}(t)$ as the equations are expressed for any slot t.

Let now $\mathbf{F}(\mathbf{n}) = \mathbb{E}[\mathbf{X}(t+1) - \mathbf{X}(t)|\mathbf{X}(t) = \mathbf{n}]$ be the expected change in $\mathbf{X}(t)$ over one time slot, given that the system is at state $\mathbf{n}$. Function $\mathbf{F}(\cdot)$ is called the drift of the system, and is given by

$$F_i(\mathbf{n}) = \begin{cases} \sum_{k=1}^{m-1} n_k \tau_k (1 - \gamma_k) - n_0 \tau_0 \gamma_0 - n_0 \beta_0, & i = 0, \\ n_{i-1}(\tau_{i-1}\gamma_{i-1} + \beta_{i-1}) - n_i(\tau_i + \beta_i), & 0 < i < m - 1, \\ n_{m-2}(\tau_{m-2}\gamma_{m-2} + \beta_{m-2}) - n_{m-1}\tau_{m-1}(1 - \gamma_{m-1}), & i = m - 1. \end{cases} \tag{6.23}$$

Equation (6.23) is obtained by balancing, for every backoff stage, the average number of stations that enter and leave this backoff stage. For example, n_0 increases by 1 only when some station transmits successfully. Since such a station could be in any of the other backoff stages and there are n_k stations in stage k, this occurs with probability $\sum_{k=1}^{m-1} n_k \tau_k (1 - \gamma_k)$. Similarly, n_0 decreases when some stations at stage 0 are either involved in a collision (which occurs with probability $n_0 \tau_0 \gamma_0$), or do not transmit and sense the medium busy $d_0 + 1$ times (which occurs with probability $n_0 \beta_0$). The decrease of n_0 in both cases is 1, thus the expected decrease is equal to the sum of the two probabilities. The resulting drift F_0 is computed by adding all these (positive and negative) expected changes in n_0.

F_i, $0 < i < m - 1$ is computed by observing that in these backoff stages, n_i decreases if and only if some stations at stage i sense the medium busy or transmit; n_i increases if and only if some stations at stage $i - 1$ sense the medium busy or transmit and collide. Finally, n_{m-1} increases if any stations at stage $m - 2$ experience a collision or sense the medium busy $d_{m-2} + 1$ times. It decreases only after a successful transmission at stage $m - 1$.

The evolution of the expected number of stations $\bar{\mathbf{n}}(t) \doteq \mathbb{E}[\mathbf{X}(t)]$ is described by the m-dimensional dynamical system

$$\bar{\mathbf{n}}(t + 1) = \bar{\mathbf{n}}(t) + \mathbf{F}(\bar{\mathbf{n}}(t)), \tag{6.24}$$

where $\mathbf{F}(\bar{\mathbf{n}}(t))$ is given by Equation (6.23). We name this model after Equation (6.23), as the Drift model.

6.8.2.2 Steady State Analysis of the System

We can compute the equilibrium point(s) of system (6.24), which is the stationary regime where the average number of stations at each backoff stage remains constant. This information enables us later to compute the normalized throughput of the network. To compute the equilibrium point(s) of (6.24), we impose the condition $\mathbf{F}(\bar{\mathbf{n}}) = \mathbf{0}$, which yields

$$\bar{n}_i = \left(\frac{\tau_{i-1}\gamma_{i-1} + \beta_{i-1}}{\tau_i + \beta_i}\right)\bar{n}_{i-1}, \ 1 \le i \le m-2, \quad \bar{n}_{m-1} = \left(\frac{\tau_{m-2}\gamma_{m-2} + \beta_{m-2}}{\tau_{m-1}(1 - \gamma_{m-1})}\right)\bar{n}_{m-2}. \quad (6.25)$$

Let us define

$$K_0 \doteq 1, \quad K_i \doteq \frac{\tau_{i-1}\gamma_{i-1} + \beta_{i-1}}{\tau_i + \beta_i}, \ 1 \le i \le m-2, \quad K_{m-1} \doteq \frac{\tau_{m-2}\gamma_{m-2} + \beta_{m-2}}{\tau_{m-1}(1 - \gamma_{m-1})}. \quad (6.26)$$

Since $\sum_{i=0}^{m-1} \bar{n}_i = N$, the equilibrium $\hat{\mathbf{n}}$ of (6.24) is given by

$$\hat{n}_i = \frac{N \prod_{j=0}^{i} K_j}{\sum_{k=0}^{m-1} \prod_{j=0}^{k} K_j}, \ 0 \le i \le m-1. \quad (6.27)$$

Recall that τ_i and β_i are functions of γ_i, given by (6.21) and (6.22). Thus, the $\hat{n}_i$s in (6.27) are also functions of γ_i, $0 \le i \le m-1$. From the above, substituting (6.27) in (6.18) yields a system of m equations with m unknowns γ_i for $0 \le i \le m-1$. The nonlinear system of questions can be solved with standard nonlinear solvers, such as Matlab and python.

6.8.3 Throughput Evaluation

By solving the m equations (6.27) that give the steady state number of nodes $\hat{n}_0, \ldots, \hat{n}_{m-1}$ at each backoff stage, we can find the values γ_i, $0 \le i \le m-1$, and using Equation (6.18), we can compute the values τ_i. We then compute the throughput of the network as follows. The probability that a slot is idle is p_e and is given by Equation (6.17). The probability of a successful transmission of a station at stage i is $\tau_i(1 - \gamma_i)$. Therefore, the probability p_s that a slot contains a successful transmission is given by $p_s = \sum_{i=0}^{m-1} \hat{n}_i \tau_i(1 - \gamma_i)$. Let p_c denote the probability that a slot contains a collision. We have $p_c = 1 - p_e - p_s$. We now have enough information to compute the normalized throughput S of the network as

$$S = \frac{p_s D}{p_s T_s + p_c T_c + p_e \sigma}, \quad (6.28)$$

where D is the frame duration, T_s is the duration of a successful transmission, T_c is the duration of a collision, and σ is the time slot duration given in Table 6.1.

6.8.4   Uniqueness of the Equilibrium Point

Similarly to the fixed-point equation (6.10), we need to ensure that this model has a unique equilibrium point given by (6.27). One sufficient condition for the uniqueness of the solution is that the sequence τ_i is decreasing with i for any n_i distribution. Similar studies for the Wi-Fi MAC protocol [4, 17] require the same sufficient condition (i.e., τ_i decreasing with i) for the model to admit a unique solution. We formally define this condition as follows:

$$\tau_i > \tau_{i+1}, \; 0 \le i \le m - 2. \tag{6.29}$$

Theorem 6.2 states that if (6.29) is satisfied, the equilibrium point given by (6.27) is unique. The proof is omitted, as it follows similar methodology to the model introduced in Section 6.2, and it is beyond the scope of this book.

THEOREM 6.2. *If (6.29) is satisfied, then the system of equations formed by (6.27) and (6.18) for $0 \le i \le m - 1$ has a unique solution.*

6.8.4.1   **Protocol Configurations Satisfying** (6.29)

Before showing in Theorem 6.3 that (6.29) is satisfied for a wide range of configurations, we introduce a useful lemma. Note that compared to 802.11, where τ_i is a function of only CW_i, the analysis here is substantially more challenging, because τ_i is a function of CW_i, d_i, and γ_i.

We have to investigate the relationship between τ_i and τ_{i+1}. Recall that these two transmission probabilities are functions of two different collision probabilities γ_i and γ_{i+1}, respectively, which makes the analysis challenging. Assume that the collision probability is the same for two successive backoff stages i and $i + 1$ and is equal to γ_i. Under this hypothesis, in Lemma 6.3, we show that if $\tau_i(\gamma_i) > \tau_{i+1}(\gamma_i)$, $\forall \gamma_i \in [0, 1]$, then $\tau_i(\gamma_i) > \tau_{i+1}(\gamma_{i+1})$, for any pair γ_i, γ_{i+1} that satisfies (6.18). In Theorem 6.3, we provide some sufficient conditions to guarantee that $\tau_i(\gamma_i) > \tau_{i+1}(\gamma_i)$ is satisfied for all $\gamma_i \in [0, 1]$ and $0 \le i < m - 1$; from Lemma 6.3, this implies that (6.29) is satisfied. The detailed proofs can be found in [18].

LEMMA 6.3. *Let γ_i^s be a value of the collision probability at stage i. Then, if $\tau_i(\gamma_i^s) > \tau_{i+1}(\gamma_i^s)$ for all $\gamma_i^s \in [0, 1]$, we have $\tau_i(\gamma_i) > \tau_{i+1}(\gamma_{i+1})$ for any n_i distribution.*

The following theorem provides some sufficient conditions on the (CW_i, d_i) configurations that ensure that (6.29) holds. Lemma 6.3 could be employed to show that (6.29) holds for more configurations than the ones covered by the theorem. It is sufficient to show that the configuration satisfies the hypothesis of Lemma 6.3 for all $0 \le i \le m - 2$.

THEOREM 6.3. *Equation (6.29) holds if the following condition is satisfied for $0 \le i \le m - 2$:*

$$CW_{i+1} > \begin{cases} CW_i, & \text{if } d_{i+1} = d_i \\ 2CW_i - d_i - 1, & \text{otherwise.} \end{cases} \tag{6.30}$$

6.9   Drift versus Decoupling Assumption Model

The Drift and decoupling assumption (DA) models can have different accuracy depending on the network scenario. We compare the Drift model with the DA model for various configurations and numbers of stations. Figure 6.23 shows throughput with the default parameters for the two priority classes CA1 and CA3 (CA0 and CA2 are equivalent) of the 1901 standard. It compares simulation to the two models. The DA model is less accurate for CA1 when N is small, because the class CA1 uses larger contention windows, which increases the time spent in backoff and, as a result, the coupling between stations.

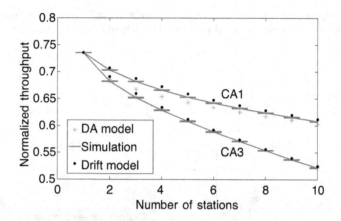

Figure 6.23 Throughput obtained by simulation and with the Drift and DA models for the default configurations of 1901 given in Table 3.4.

We study the accuracy of the two models in more general settings. We introduce a factor f such that at each stage i, the value of d_i is given by $d_i = f^i(d_0 + 1) - 1$. This enables us to define various sequences of values for the d_is, using only f and d_0. At each stage i, CW_i is given by $CW_i = 2^i CW_{\min}$, and there are m backoff stages ($i \in \{0, m - 1\}$). In Figure 6.24, we show the throughput for various such values of d_0 and f, with $CW_{\min} = 8$ and $m = 5$. We observe that the DA model achieves good accuracy when the d_is are large, because in these configurations, the deferral counter is less likely to expire, which reduces the coupling among stations. The drift model achieves good accuracy when the d_is are small, while there is a small deviation for large d_is; this is due to the assumptions of our network model of Section 6.8.2.1, which are not used by the DA model.[1]

In Figure 6.25, we evaluate the accuracy of the Drift model on the collision probability at each backoff stage i of the CA1 class of IEEE 1901. The Drift model can accurately estimate the collision probabilities γ_i, $0 \leq i \leq 3$.

We summarize the differences between the two PLC CSMA/CA models in Table 6.4.

---

[1] The Wi-Fi model that does not rely on the decoupling assumption has a similar deviation compared to Bianchi's model [1, 19].

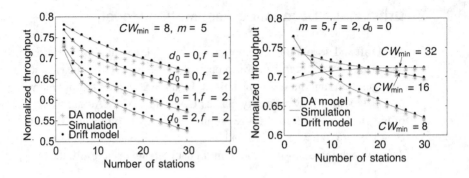

Figure 6.24 Throughput obtained by simulation, and with the Drift and DA models for different configurations. The initial values d_i of the deferral counter at each backoff stage are given by $d_i = f^i(d_0 + 1) - 1$.

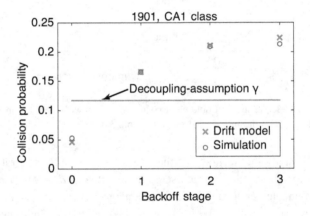

Figure 6.25 Simulations and model evaluations of the collision probability with 1901 (with CA1 parameters) for $N = 2$. Drift model accurately predicts the collision probability at each backoff stage i, γ_i. For reference, we also show the collision probability γ predicted by the decoupling assumption model.

In addition to the differences described in the table, let us note that the models have the same accuracy for large N (typically $N \geq 15$). Both models can have practical benefits and disadvantages, complementing each other. They can be used for different configurations and modeling insights.

Table 6.4 Comparison of the Drift and decoupling assumption models

Drift model	Decoupling assumption model
Coupled model of all stations	Model of a tagged station given independence
System of m nonlinear equations (complex)	Nonlinear fixed-point equation (simple)
Both transient and stationary regimes	Only stationary regime
Enables short-term dynamics analysis	Only long-term average analysis
More accurate for small N, except when d_is are large	Much less accurate for small N, except when d_is are large
Application: accurate throughput prediction for tuning the trade-off between throughput and fairness	Application: throughput enhancements for large N, thanks to simplicity

6.10   Summary

The MAC layer of PLC can react to contention proactively and without involving collisions, which can offer the potential of high gains in terms of throughput as compared to Wi-Fi. In this chapter, we have introduced performance models of PLC and studied different multiuser scenarios. One of the main challenges of modeling PLC is the complexity of the protocol that has a very large state space. To reduce the state space, we have made the assumption that the backoff processes of the stations are independent. The model performs well in comparison with testbed and simulation experiments.

We have presented results and guidelines of a cross-layer performance on PLC. We find that a rate anomaly does not exist in PLC, hence slow stations do not reduce the overall throughput. However, PLC suffers from hidden terminals, as commercial PLC devices do not use RTS/CTS.

We have studied the fairness characteristics of PLC and given a comparison with Wi-Fi. Wi-Fi is conservative and employs large contention-window sizes to avoid collisions, whereas 1901 is more aggressive: it tries small contention-window values first, but later engages a complex mechanism with the deferral counter to circumvent the high collision rate and to increase throughput. This complexity of PLC can affect fairness and delay. We have shown analytically that 1901 can be less fair than 802.11 when $N = 2$. The experiments show that the default 1901 configuration can harm delay-sensitive traffic, such as voice or video, especially when the default settings are used with priority class CA1 of the IEEE 1901 standard. End user experience depends on the existence of proper traffic-classification mechanisms that assign priorities to traffic flows.

We have also discussed a commonly adopted decoupling hypothesis for modeling the MAC layer. Although this assumption is proven analytically and experimentally to be accurate for 802.11, we have found that for 1901, it can be stronger and does not necessarily hold when there are few stations contending for the medium. We have attributed the inaccuracy of the hypothesis to the strong dependence between the PLC stations, which is also related to short-term fairness. We have presented an alternative, more complex model that does not rely on decoupling assumptions.

Appendixes

A   Decoupling Assumption Model

The proof of Theorem 6.1 is originally published in [18] and is built on a few lemmas and corollaries, which we state first.

LEMMA 6.4. *Let γ be the collision probability. Let $B_i(\gamma)$ be the expected number of backoff slots between two transmissions attempts of a station that always remains at backoff stage i. Then, we have $B_i = bc_i/t_i - 1$, and B_i is an increasing function of γ, for any $0 \le i \le m - 1$.*

Proof By its definition and by using the same rationale as the one employed to compute bc_i in (6.2), B_i is given recursively by

$$B_i = \frac{d_i(d_i + 1)}{2CW_i} + \sum_{j=d_i+1}^{CW_i-1} \frac{j(1 - x_j^i) + \sum_{k=d_i+1}^{j} (k + B_i)(x_k^i - x_{k-1}^i)}{CW_i}. \tag{6.31}$$

Now, solving (6.31) over B_i, gives $B_i = bc_i/t_i - 1$, with bc_i and t_i given by (6.2) and (6.3).

To prove the second part of the lemma, we proceed as follows. (1) First, we compute $dB_i/d\gamma$. (2) Second, we show that this derivative is positive at $\gamma = 1$. (3) Third, we show that if the derivative is negative at some $0 < \gamma^* < 1$, it will also be negative at any value $\gamma > \gamma^*$. The proof then follows by contradiction: if the derivative was negative at some γ^*, it would also be negative at $\gamma = 1$, which would contradict (2).

(1) After rearranging terms, (6.31) can be rewritten as

$$B_i = \frac{CW_i - 1}{2} + \frac{1}{CW_i} \sum_{j=d_i+1}^{CW_i-1} \left(B_i x_j^i - \sum_{k=d_i+1}^{j-1} x_k^i \right). \tag{6.32}$$

The derivative of B_i can be computed as

$$\frac{dB_i}{d\gamma} = \sum_{k=d_i+1}^{CW_i-1} \frac{\partial B_i}{\partial x_k^i} \frac{dx_k^i}{d\gamma}. \tag{6.33}$$

The partial derivative $\partial B_i/\partial x_k^i$ can be computed from (6.32) as

$$\frac{\partial B_i}{\partial x_k^i} = \frac{B_i - (CW_i - 1 - k)}{CW_i} + \frac{\partial B_i}{\partial x_k^i} \sum_{j=d_i+1}^{CW_i-1} \frac{x_j^i}{CW_i}, \tag{6.34}$$

which yields

$$\frac{dB_i}{d\gamma} = \frac{\sum_{k=d_i+1}^{CW_i-1} (B_i - (CW_i - 1 - k)) \frac{dx_k^i}{d\gamma}}{CW_i - \sum_{j=d_i+1}^{CW_i-1} x_j^i}. \tag{6.35}$$

To compute $dx_k^i/d\gamma$, we observe that x_k^i is the complementary cumulative function of a binomial distribution. By taking its derivative, we obtain

$$\frac{dx_k^i}{d\gamma} = \frac{k!}{(k - d_i - 1)!d_i!}\gamma^{d_i}(1 - \gamma)^{k-d_i-1}. \tag{6.36}$$

(2) Next, we show that $dB_i/d\gamma > 0$ at $\gamma = 1$. We have $x_k^i = 1$ at $\gamma = 1$ for all $d_i + 1 \leq k \leq CW_i - 1$. Given this and (6.31), we have

$$B_i = \frac{d_i(d_i + 1)}{2CW_i} + \frac{CW_i - d_i - 1}{CW_i}(d_i + 1 + B_i). \tag{6.37}$$

Solving (6.37) over B_i yields $B_i = CW_i - d_i/2 - 1$. Now, notice that $dx_k^i/d\gamma = 0$ at $\gamma = 1$ for all $d_i + 1 < k \leq CW_i - 1$, and $dx_{d_i+1}^i/d\gamma = d_i + 1$ from (6.36). Substituting in (6.35) yields $dB_i/d\gamma = d_i/2 + 1$, i.e., $dB_i/d\gamma > 0$.

(3) Let us now assume that $dB_i/d\gamma < 0$ for some $\gamma^* < 1$. Let $l = \lceil CW_i - 1 - B_i(\gamma^*)\rceil$. Given (6.35), we can express $dB_i/d\gamma$ as the product of two terms, $dB_i/d\gamma = f_1(\gamma)f_2(\gamma)$, where

$$f_1(\gamma) \doteq \frac{dx_l^i/d\gamma}{CW_i - \sum_{j=d_i+1}^{CW_i-1} x_j^i}, \quad f_2(\gamma) \doteq \sum_{k=d_i+1}^{CW_i-1}(B_i - (CW_i - 1 - k))\frac{dx_k^i/d\gamma}{dx_l^i/d\gamma}. \tag{6.38}$$

We have $f_1(\gamma) > 0 \; \forall\gamma$, which implies $dB_i/d\gamma < 0$ if and only if $f_2(\gamma) < 0$. Also, we have

$$\frac{df_2(\gamma)}{d\gamma} = \sum_{k=d_i+1}^{CW_i-1}\frac{dB_i}{d\gamma}\frac{dx_k^i/d\gamma}{dx_l^i/d\gamma} + \sum_{k=d_i+1}^{l-1}(B_i - (CW_i - 1 - k))\frac{d}{d\gamma}\left(\frac{dx_k^i/d\gamma}{dx_l^i/d\gamma}\right) \tag{6.39}$$

$$+ \sum_{k=l+1}^{CW_i-1}(B_i - (CW_i - 1 - k))\frac{d}{d\gamma}\left(\frac{dx_k^i/d\gamma}{dx_l^i/d\gamma}\right),$$

and

$$\frac{d}{d\gamma}\left(\frac{dx_k^i/d\gamma}{dx_l^i/d\gamma}\right) = -\frac{k!(l - d_i - 1)!}{l!(k - d_i - 1)!}(k - l)(1 - \gamma)^{k-l-1}, \tag{6.40}$$

which is positive for $k < l$ and negative for $k > l$. From the above equations, it follows that as long as $CW_i - 1 - (l - 1) > B_i(\gamma) > CW_i - 1 - l$ and $dB_i/d\gamma < 0$, we have $df_2/d\gamma < 0$.

Building on the above, next we show that $B_i(\gamma)$ decreases for $\gamma \in [\gamma^*, \gamma^l]$, where γ^l is the γ value for which $B_i(\gamma^l) - (CW_i - 1 - l) = 0$. At $\gamma = \gamma^*$, we have $f_2(\gamma^*) < 0$, $dB_i/d\gamma < 0$, and $df_2/d\gamma < 0$. Let us assume that, before $B_i(\gamma)$ decreases down to $CW_i - 1 - l$, there is some $\hat{\gamma} > \gamma^*$ for which $dB_i/d\gamma \geq 0$. This implies that for some $\gamma' \in (\gamma^*, \hat{\gamma})$, $f_2(\gamma)$ has to stop decreasing, i.e., $df_2(\gamma')/d\gamma = 0$. Since $f_2(\gamma)$ decreases in $[\gamma^*, \gamma']$, we have $f_2(\gamma) < 0$ for $\gamma \in [\gamma^*, \gamma']$. Thus, $B_i(\gamma)$ decreases in $[\gamma^*, \gamma']$. As $CW_i - 1 - (l - 1) > B_i(\gamma^*) > CW_i - 1 - l$ and (by assumption) $B_i(\gamma)$ does not reach $CW_i - 1 - l$, we also have $CW_i - 1 - (l - 1) > B_i(\gamma') > CW_i - 1 - l$, which contradicts $df_2(\gamma')/d\gamma = 0$. Hence, our assumption does not hold, and $dB_i/d\gamma < 0$ until B_i reaches $CW_i - 1 - l$, i.e., $dB_i/d\gamma < 0$ for $\gamma \in [\gamma^*, \gamma^l]$.

Following the same rationale for $\gamma \in [\gamma^l, \gamma^{l+1}]$, we can prove that $dB_i/d\gamma < 0$ for $\gamma \in [\gamma^l, \gamma^{l+1}]$. We can repeat this recursively to show that $dB_i/d\gamma < 0$ for $\gamma \in [\gamma^{l+1}, \gamma^{l+2}]$, $\gamma \in [\gamma^{l+2}, \gamma^{l+3}]$ until reaching $\gamma = 1$, which yields a contradiction because $dB_i/d\gamma > 0$ at $\gamma = 1$ from step (2) above. $\square$

COROLLARY 6.1. *If* $CW_{i+1} \geq 2CW_i - d_i - 1$, *then we have* $B_{i+1} > B_i$, *for any* $0 \leq i \leq m-2$.

Proof By Lemma 6.4, the minimum value of B_{i+1} is $B_{i+1}^{min} = (CW_{i+1} - 1)/2$ at $\gamma = 0$, and the maximum value of B_i is $B_i^{max} = CW_i - d_i/2 - 1$ at $\gamma = 1$. Setting $CW_{i+1} \geq 2CW_i - d_i - 1$, yields $B_{i+1}^{min} \geq B_i^{max}$, hence $B_{i+1} > B_i$ for all $\gamma \in [0, 1]$. $\square$

COROLLARY 6.2. *If* $CW_{i+1} = CW_i$ *and* $d_{i+1} < d_i$, *then we have* $B_{i+1} \geq B_i$, *for any* $0 \leq i \leq m - 2$.

Proof By the proof of Lemma 6.4, the equality holds for $\gamma = 0$, because $B_i(0) = (CW_i + 1)/2$. We now show that for $\gamma \in (0, 1]$, we have $B_{i+1} > B_i$.

If we prove $B_{i+1} > B_i$ when $CW_{i+1} = CW_i$, $d_{i+1} = d$ and $d_i = d + 1$, then the corollary follows by induction. Thus, we now show $B_{i+1} > B_i$ for this case, and we proceed as follows. Given (6.32), the difference $B_{i+1} - B_i$ can be computed as

$$B_{i+1} - B_i = \frac{\sum_{j=d+1}^{CW_i-1} \left(B_{i+1}x_j^{i+1} - \sum_{k=d+1}^{j-1} x_k^{i+1} \right)}{CW_i} - \frac{\sum_{j=d+2}^{CW_i-1} \left(B_i x_j^i - \sum_{k=d+2}^{j-1} x_k^i \right)}{CW_i}. \quad (6.41)$$

We have $x_j^{i+1} = x_j^i + \binom{j}{d+1}\gamma^{d+1}(1-\gamma)^{j-d-1}$. Let $\delta_j \doteq \binom{j}{d+1}\gamma^{d+1}(1-\gamma)^{j-d-1}$. Then, we have $\delta_j = x_j^{i+1} - x_j^i$. By rearranging the terms and solving over the difference $B_{i+1} - B_i$ in (6.41), and by using the definition of δ_j, we have

$$B_{i+1} - B_i = \frac{\sum_{k=d+1}^{CW_i-1} (B_{i+1} - (CW_i - 1 - k))\delta_k}{CW_i - \sum_{j=d+2}^{CW_i-1} x_j^i}. \quad (6.42)$$

By using (6.36), we now observe that

$$\delta_j = \frac{dx_j^{i+1}}{d\gamma} \frac{\gamma}{d + 1}. \quad (6.43)$$

Substituting this in (6.42), yields

$$B_{i+1} - B_i = \frac{\sum_{j=d+1}^{CW_i-1} (B_{i+1} - (CW_i - 1 - j))\frac{\gamma}{d+1}\frac{dx_j^{i+1}}{d\gamma}}{CW_i - \sum_{j=d+2}^{CW_i-1} x_j^i} \quad (6.44)$$

$$= \frac{CW_i - \sum_{j=d+1}^{CW_i-1} x_j^{i+1}}{CW_i - \sum_{j=d+2}^{CW_i-1} x_j^i} \frac{\gamma}{d + 1} \left(\frac{dB_{i+1}}{d\gamma} \right) > 0, \quad (6.45)$$

where the inequality holds by Lemma 6.4 and for all $\gamma \in (0, 1]$. $\square$

COROLLARY 6.3. *If* $CW_{i+1} > CW_i$ *and* $d_{i+1} = d_i$, *then we have* $B_{i+1} > B_i$, *for any* $0 \leq i \leq m - 2$.

Proof If $B_{i+1} > B_i$ holds for $CW_{i+1} = CW_i + 1$, by using induction it is easy to see that it holds for any $CW_{i+1} > CW_i$. Thus, we prove the corollary for $CW_{i+1} = CW_i + 1$.

Because $d_i = d_{i+1}$ and by using (6.1), we have $x_k^i = x_k^{i+1}$ for all $d_i + 1 \le k \le CW_i - 1$. Given this, we have

$$B_{i+1} - B_i = \frac{(CW_i - \sum_{j=d_i+1}^{CW_i-1} x_j^i)^2}{(CW_i + 1 - \sum_{j=d_i+1}^{CW_i} x_j^i)(CW_i - \sum_{j=d_i+1}^{CW_i-1} x_j^i)} \quad (6.46)$$
$$- \frac{(1 - x_{CW_i}^i)^{\frac{CW_i(CW_i-1)}{2}}}{(CW_i + 1 - \sum_{j=d_i+1}^{CW_i} x_j^i)(CW_i - \sum_{j=d_i+1}^{CW_i-1} x_j^i)}$$
$$+ \frac{(1 - x_{CW_i}^i)\sum_{j=d_i+1}^{CW_i-1}(CW_i - 1 - j)x_j^i}{(CW_i + 1 - \sum_{j=d_i+1}^{CW_i} x_j^i)(CW_i - \sum_{j=d_i+1}^{CW_i-1} x_j^i)}.$$

By the definition of x_k^i in (6.1), we have $x_{CW_i}^i \ge x_k^i$, $d_i + 1 \le k \le CW_i - 1$, with equality at $\gamma = 0, 1$. This yields $\sum_{j=d_i+1}^{CW_i-1} x_j^i \le (CW_i - d_i - 1)x_{CW_i}^i < CW_i x_{CW_i}^i$. We thus have

$$B_{i+1} - B_i > \frac{CW_i(1 - x_{CW_i}^i)(CW_i - \sum_{j=d_i+1}^{CW_i-1} x_j^i)}{(CW_i + 1 - \sum_{j=d_i+1}^{CW_i} x_j^i)(CW_i - \sum_{j=d_i+1}^{CW_i-1} x_j^i)} \quad (6.47)$$
$$- \frac{(1 - x_{CW_i}^i)^{\frac{CW_i(CW_i-1)}{2}}}{(CW_i + 1 - \sum_{j=d_i+1}^{CW_i} x_j^i)(CW_i - \sum_{j=d_i+1}^{CW_i-1} x_j^i)}$$
$$+ \frac{(1 - x_{CW_i}^i)\sum_{j=d_i+1}^{CW_i-1}(CW_i - 1 - j)x_j^i}{(CW_i + 1 - \sum_{j=d_i+1}^{CW_i} x_j^i)(CW_i - \sum_{j=d_i+1}^{CW_i-1} x_j^i)}$$
$$= \frac{(1 - x_{CW_i}^i)(\frac{CW_i(CW_i+1)}{2} - \sum_{j=d_i+1}^{CW_i-1}(j + 1)x_j^i)}{(CW_i + 1 - \sum_{j=d_i+1}^{CW_i} x_j^i)(CW_i - \sum_{j=d_i+1}^{CW_i-1} x_j^i)} \ge 0,$$

where the last inequality holds because $x_j^i \le 1$, for all i, j. □

We now present the proof of Lemmas 6.1 and 6.2, and Theorem 6.1, introduced in Section 6.2.

Proof of Lemma 6.1 Let X_i be the random variable denoting the number of slots that a station that starts in stage i spends in backoff before transmitting its current packet successfully. With this notation, it holds $\mathbb{E}[X] = \mathbb{E}[X_0]$. Let c_i and j_i denote the probabilities that a station at stage i ends this stage due to a collision, or due to sensing the medium busy $d_i + 1$ times, respectively. Note that $s_i + c_i + j_i = 1$ for all i. Additionally, let bc_{s_i}, bc_{c_i}, and bc_{j_i} be the expected number of backoff slots that a station spends in backoff stage i, given that the station ends up redrawing its backoff counter due to a packet successfully transmitted, due to a collision, or due to sensing the medium busy, respectively. From the law of total probability, we have

$$\mathbb{E}[X] = s_0 bc_{s_0} + c_0 bc_{c_0} + j_0 bc_{j_0} + (c_0 + j_0)\mathbb{E}[X_1] = bc_0 + (1 - s_0)\mathbb{E}[X_1]. \quad (6.48)$$

Repeating the above reasoning recursively for $\mathbb{E}[X_1], \mathbb{E}[X_2], \ldots, \mathbb{E}[X_{m-1}]$, we have

$$\mathbb{E}[X_0] = bc_0 + (1 - s_0)\Big(bc_1 + (1 - s_1)(bc_2 + \ldots (1 - s_{m-2})\mathbb{E}[X_{m-1}])\Big). \quad (6.49)$$

Now, by applying the same reasoning as in (6.48), we have

$$\mathbb{E}[X_{m-1}] = bc_{m-1} + (1 - s_{m-1})\mathbb{E}[X_{m-1}]. \quad (6.50)$$

Solving for $\mathbb{E}[X_{m-1}]$, we obtain $\mathbb{E}[X_{m-1}] = bc_{m-1}/s_{m-1}$. Plugging this expression into (6.49) concludes the proof. □

Proof of Lemma 6.2   Similar to the proof of Lemma 6.1. □

Proof of Theorem 6.1   By Brouwer's fixed-point theorem, since $\Gamma(G(\gamma))$ is a continuous function, there exists a fixed point in $[0, 1]$. Furthermore, if $\Gamma(G(\gamma))$ is monotone, this fixed point is unique. As $\Gamma(\tau) = 1 - (1 - \tau)^{N-1}$ is nondecreasing in γ, it is thus sufficient to show that $G(\gamma)$ is monotone in γ.

Let $Q(\gamma) = (1 - \gamma)\mathbb{E}[X]$. Then, we have $G(\gamma) = 1/Q(\gamma)$. Now, $G(\gamma)$ is nonincreasing in γ if and only if $Q(\gamma)$ is nondecreasing in γ, which we show in the following.

We now use Lemma 6.4 to express $Q(\gamma)$ as a function of B_i. Replacing bc_i with $t_i(B_i + 1)$ in the expression for $\mathbb{E}[X]$ and using $s_i = (1 - \gamma)t_i$, $Q(\gamma)$ can be rewritten as

$$Q(\gamma) = \sum_{i=0}^{m-2} s_i(B_i + 1) \prod_{j=0}^{i-1}(1 - s_j) + \prod_{i=0}^{m-2}(1 - s_i)(B_{m-1} + 1). \quad (6.51)$$

The derivative of $Q(\gamma)$ with respect to γ is given by

$$\frac{dQ}{d\gamma} = \sum_{i=0}^{m-2} s_i \frac{dB_i}{d\gamma} \prod_{j=0}^{i-1}(1 - s_j) + \prod_{i=0}^{m-2}(1 - s_i)\frac{dB_{m-1}}{d\gamma} \quad (6.52)$$

$$- \sum_{i=0}^{m-2} \frac{ds_i}{d\gamma}\left[\sum_{j=i+1}^{m-2}\left((B_j + 1)s_j \frac{\prod_{k=0}^{j-1}(1 - s_k)}{1 - s_i}\right)\right.$$

$$\left. - (B_i + 1)\prod_{j=0}^{i-1}(1 - s_j) + \frac{\prod_{j=0}^{m-2}(1 - s_j)}{1 - s_i}(B_{m-1} + 1)\right].$$

From Lemma 6.4, we have that $dB_i/d\gamma > 0$. Thus, the first two terms in (6.52) are positive and it follows that

$$\frac{dQ}{d\gamma} > -\sum_{i=0}^{m-2}\frac{ds_i}{d\gamma}\left[\sum_{j=i+1}^{m-2}\left((B_j + 1)s_j \frac{\prod_{k=0}^{j-1}(1 - s_k)}{1 - s_i}\right)\right. \quad (6.53)$$

$$\left. - (B_i + 1)\prod_{j=0}^{i-1}(1 - s_j) + \frac{\prod_{j=0}^{m-2}(1 - s_j)}{1 - s_i}(B_{m-1} + 1)\right].$$

In the proof of Lemma 6.4, we show that x_k^i is increasing with γ (see (6.36)). Thus, s_i is decreasing with γ, since, from (6.3), it holds

$$\frac{ds_i}{d\gamma} = -t_i - (1 - \gamma)\frac{\sum_{k=d_i+1}^{CW_i-1} dx_k^i/d\gamma}{CW_i} < 0. \quad (6.54)$$

From Corollaries 1–3 and Condition (6.11) of Theorem 6.1, $B_i(\gamma)$ is nondecreasing with i. Combining these two properties (i.e., $ds_i/d\gamma < 0$, $B_{i+1} \geq B_i$) for $i = m - 2$ with (6.53), we have

$$\frac{dQ}{d\gamma} \geq -\sum_{i=0}^{m-3} \frac{ds_i}{d\gamma} \frac{1}{1-s_i} \left[\sum_{j=i+1}^{m-2} \left((B_j+1)s_j \prod_{k=0}^{j-1}(1-s_k) \right) \right.$$

$$\left. - (B_i+1) \prod_{j=0}^{i}(1-s_j) + \prod_{j=0}^{m-2}(1-s_j)(B_{m-1}+1) \right]. \qquad (6.55)$$

By using $ds_i/d\gamma < 0$, $B_{i+1} \geq B_i$ in the above inequality and rearranging the factors in products involving the s_is, we have

$$\frac{dQ}{d\gamma} \geq -\sum_{i=0}^{m-3} \frac{ds_i}{d\gamma} \frac{1}{1-s_i} \left[(B_i+1) \sum_{j=i+1}^{m-2} \left(s_j \prod_{k=0}^{j-1}(1-s_k) \right) \right.$$

$$\left. - (B_i+1) \prod_{j=0}^{i}(1-s_j) + \prod_{j=0}^{m-2}(1-s_j)(B_{m-1}+1) \right]$$

$$= -\sum_{i=0}^{m-3} \frac{ds_i}{d\gamma} \frac{\prod_{j=0}^{m-2}(1-s_j)}{1-s_i}(B_{m-1} - B_i) \geq 0, \qquad (6.56)$$

with equality at $\gamma = 0$. This completes the proof. $\qquad\qquad\qquad\qquad\qquad$ □

B   Distribution of Intertransmissions

The proof is originally published in [18]. We present the computation of the distribution of intertransmissions for IEEE 1901, given in Proposition 6.1 of Section 6.4.2. We assume that the backoff counters are continuous random variables, thus there are no collisions between the stations. The station with the smallest backoff counter wins the channel whenever the stations resume or choose their backoff counters. Thus, the distribution of the number of intertransmissions K, $\mathbb{P}(K = k)$, is computed by integrating the probability density functions of M backoff counters over the appropriate region of $\mathbb{R}^M$ (M depends on k as shown in the next paragraphs). We next prove the results for CA0/CA1 priorities (see Table 3.4). The distribution is computed similarly for CA2/CA3 priorities.

We assume that we have two stations A and B and that both stations start at the same state, i.e., backoff stage 0. We tag one of the stations in order to compute its distribution of K. Let B be the tagged station and A be the nontagged station. We denote by $b^{(i)}$ the backoff counter that B draws when it enters backoff stage i. Thus, $b^{(0)}$, $b^{(1)}$, $b^{(2)}$, $b^{(3)}$ are uniformly distributed over the ranges $[0, 8]$, $[0, 16]$, $[0, 32]$, $[0, 64]$, respectively. After reaching the maximum backoff stage which is 3, the station does not change the contention-window value. Thus, $b^{(i)}$ are uniformly distributed over the range $[0, 64]$ for $i \geq 3$, and B chooses a new backoff counter $b^{(i)}$ after every 16 consecutive transmissions of A, due to the deferral-counter value at this backoff stage (which is 15). We denote by

Table 6.5 The deferral counter of B, DC_B, and the backoff counter of B, BC_B, before the kth transmission of A

k	DC_B	BC_B
0	0	$b^{(0)}$
1	0	$b^{(0)}$
2	1	$b^{(1)}$
3	0	$b^{(1)} - a_2$
4	3	$b^{(2)}$
5	2	$b^{(2)} - a_4$
6	1	$b^{(2)} - a_4 - a_5$
7	0	$b^{(2)} - a_4 - a_5 - a_6$
8	15	$b^{(3)}$
9	14	$b^{(3)} - a_8$
$\vdots$	$\vdots$	$\vdots$
23	0	$b^{(3)} - a_8 - \sum_{i=8}^{22} a_i$
24	15	$b^{(4)}$

a_k the backoff counter of A before the kth consecutive transmission, and a_k is uniformly distributed over the range $[0, 8]$, for all $k \geq 0$.

Due to the definition of 1901 protocol and to the continuous-time backoff counters assumption, stations change a backoff stage only when they sense the medium busy and their deferral counter is 0. As A is the one that transmits successfully before B wins the channel, A is always at backoff stage 0. B senses the medium busy and depending on the deferral-counter value after k consecutive transmissions of A, B chooses a new backoff counter. All a_k are mutually independent and independent of all $b^{(i)}$, because in the real protocol stations choose backoff counters independently of the other stations and of the backoff counters the stations selected in the past.

We normalize the distributions of the backoff counters, so that all a_k and $b^{(0)}$ are uniformly distributed over the range $[0, 1]$. Hence, $b^{(1)}$, $b^{(2)}$, $b^{(3)}$ are uniformly distributed over the ranges $[0, 2]$, $[0, 4]$, $[0, 8]$, respectively. This shows that the distribution of K does not depend on the minimum contention-window value as long as we keep the binary exponential backoff of class CA1 of 1901. We denote by DC_B and BC_B the random variables that represent the backoff- and deferral-counter values at any state k, as shown in Table 6.5. Given the above, $\mathbb{P}(K = k)$ is computed as follows:

- For $k = 0$, B transmits, thus $\mathbb{P}(K = 0) = P(a_1 \geq b^{(0)})$.
- For $k = 1$, A transmits once and then B transmits. Because DC_B is 0 when B enters backoff stage 0 (see Table 3.4), B enters backoff stage 1 and doubles CW after sensing the medium busy once. Hence, we have

$$\mathbb{P}(K = 1) = \mathbb{P}\left(a_1 < b^{(0)} \text{ and } a_2 \geq b^{(1)}\right). \tag{6.57}$$

- For $k = 2$, A transmits twice and then B transmits, thus

$$\mathbb{P}(K = 2) = \mathbb{P}\left(a_1 < b^{(0)} \text{ and } a_2 < b^{(1)} \text{ and } a_3 \geq b^{(1)} - a_2\right), \tag{6.58}$$

because DC_B is 1 when B enters backoff stage 1.

- For $k = 3$, A transmits three times and then B transmits:

$$\mathbb{P}(K = 3) = \mathbb{P}\left(a_1 < b^{(0)} \text{ and } a_3 < b^{(1)} - a_2 \text{ and } a_4 \geq b^{(2)}\right), \tag{6.59}$$

because DC_B becomes 0 after one transmission of A, at backoff stage 1.

- For $k \in [4, 6]$, A transmits k times and then B transmits:

$$\mathbb{P}(K = k) = \mathbb{P}\left(a_1 < b^{(0)}\right) \mathbb{P}\left(a_3 + a_2 < b^{(1)}\right) \tag{6.60}$$

$$\cdot \mathbb{P}\left(a_k < b^{(2)} - \sum_{i=4}^{k-1} a_i \text{ and } a_{k+1} \geq b^{(2)} - \sum_{i=4}^{k} a_i\right), \tag{6.61}$$

due to the independence of the backoff counters. After the kth transmission of A, B's backoff counter equals $b^{(2)} - \sum_{i=4}^{k} a_i$.

- When $k = 7$, DC_B is 0. Thus, at the 8th attempt of A, B has a backoff counter $b^{(3)}$ and $DC_B = 15$. Thus,

$$\mathbb{P}(K = 7) = \mathbb{P}\left(a_1 < b^{(0)} \text{ and } a_3 + a_2 < b^{(1)}\right) \mathbb{P}\left(a_7 < b^{(2)} - \sum_{i=4}^{6} a_i \text{ and } a_8 \geq b^{(3)}\right). \tag{6.62}$$

- For $8 \leq k \leq 22$, B decrements DC_B by 1 after each successful transmission of A, and the backoff counter is $b^{(3)} - \sum_{i=8}^{k} a_i$ when A has transmitted k times. Hence,

$$\mathbb{P}(K = k) = \mathbb{P}\left(a_1 < b^{(0)}\right) \mathbb{P}\left(a_3 + a_2 < b^{(1)}\right) \mathbb{P}\left(a_7 < b^{(2)} - \sum_{i=4}^{6} a_i\right) \tag{6.63}$$

$$\cdot \mathbb{P}\left(a_k < b^{(3)} - \sum_{i=8}^{k-1} a_i \text{ and } a_{k+1} \geq b^{(3)} - \sum_{i=8}^{k} a_i\right). \tag{6.64}$$

- When $k = 23$, $DC_B = 0$, thus at the next attempt B chooses a new backoff counter $b^{(4)}$ and the new DC_B is 15. Therefore, we have

$$\mathbb{P}(K = 23) = \mathbb{P}\left(a_1 < b^{(0)}\right) \mathbb{P}\left(a_3 + a_2 < b^{(1)}\right) \tag{6.65}$$

$$\cdot \mathbb{P}\left(a_7 < b^{(2)} - \sum_{i=4}^{6} a_i\right) \mathbb{P}\left(a_{23} < b^{(3)} - \sum_{i=8}^{22} a_i\right) \mathbb{P}\left(a_{24} \geq b^{(4)}\right). \tag{6.66}$$

We now explain the computation details of the above probabilities. For $k < 15$ we integrate first with respect to the appropriate $b^{(i)}$, because the maximum value of the sum of the backoff counters of A is less or equal than the maximum value of $b^{(i)}$. For $15 \leq k \leq 22$ the maximum value of the sum of the backoff counters of A is larger than the maximum value of $b^{(i)}$. Thus, we define a new variable which is the sum of the backoff counters of A and we integrate first with respect to this random variable. Let U be the sum of n independent random variables uniformly distributed over the range $[0, 1]$. Then, U follows the Irwin-Hall distribution with probability density function

$$f_{I-H}(u; n) = \frac{1}{2(n-1)!} \sum_{k=0}^{n} (-1)^k \binom{n}{k} (u - k)^{n-1} \operatorname{sgn}(u - k). \tag{6.67}$$

From the above, for $k \geq 7$

$$P(K = k) = A_1 \begin{cases} A_2 A_3^l & 7 + 16l \leq k \leq 14 + 16l \\ \\ A_3^l \mathbf{I}(k - 16l - 14) & 15 + 16l \leq k \leq 22 + 16l, \end{cases} \tag{6.68}$$

where l is $\lfloor (k-7)/16 \rfloor$ and $A_1, A_2, A_3, \mathbf{I}$ are given by (6.69), (6.70), (6.71), (6.72) below.

$$A_1 = \mathbb{P}(a_1 < b^{(0)})\mathbb{P}(a_3 + a_2 < b^{(1)})\mathbb{P}\left(\sum_{i=4}^{7} a_i < b^{(2)}\right) = \frac{1}{2}\frac{1}{2}\frac{1}{2} = \frac{1}{8}. \tag{6.69}$$

$$A_2 = \mathbb{P}(a_8 \geq b^{(3)}) = \mathbb{P}(a_{24} \geq b^{(4)}) = \frac{1}{16}. \tag{6.70}$$

$$A_3 = \mathbb{P}\left(\sum_{i=8}^{23} a_i < b^{(3)}\right) = \int_0^8 \frac{1}{8} db^{(3)} \int_0^{b^{(3)}} f_{I-H}(u; 16) du = 0.0578. \tag{6.71}$$

$\mathbf{I}$ is an 8-element vector. The element in position $x - 14$ ($x \in [15, 22]$) is given by

$$\mathbf{I}(x - 14) = \mathbb{P}\left(\sum_{i=8}^{x} a_i < b^{(3)} \text{ and } \sum_{i=8}^{x+1} a_i \geq b^{(3)}\right) = \int_0^1 da_{x+1} \int_0^8 \frac{1}{8} db^{(3)} \int_{b^{(3)}-a_{x+1}}^{b^{(3)}} f_{I-H}(u; x - 7) du. \tag{6.72}$$

We have shown the computations for $k \in [15, 23]$ and $l = 0$. For $24 \leq k \leq 38$ and $l = 1$, we have

$$P(K = k) = \mathbb{P}(a_1 < b^{(0)})\mathbb{P}(a_3 + a_2 < b^{(1)})\mathbb{P}\left(\sum_{i=4}^{7} a_i < b^{(2)}\right)\mathbb{P}\left(\sum_{i=8}^{23} a_i < b^{(3)}\right) \tag{6.73}$$

$$\cdot \mathbb{P}\left(\sum_{i=24}^{k} a_i < b^{(4)} \text{ and } \sum_{i=24}^{k+1} a_i \geq b^{(4)}\right), \tag{6.74}$$

due to the independence of the backoff counters. The last factor is equal to

$$\mathbb{P}\left(\sum_{i=8}^{k} a_i < b^{(3)} \text{ and } \sum_{i=8}^{k+1} a_i \geq b^{(3)}\right) \text{ for } 8 \leq k \leq 22, \tag{6.75}$$

thus the distribution for $23 \leq k \leq 38$ is equal to the distribution for $7 \leq k \leq 22$ multiplied by A_3. For $39 \leq k \leq 54$ the distribution is equal to the distribution for $7 \leq k \leq 22$ multiplied by A_3^2, and so on. This completes the proof.

References

[1] G. Bianchi, "Performance analysis of the IEEE 802.11 distributed coordination function," *IEEE Journal on Selected Areas in Communications (JSAC)*, vol. 18, no. 3, pp. 535–547, 2000.

[2] M. Chung, M. Jung, T. Lee, and Y. Lee, "Performance analysis of HomePlug 1.0 MAC with CSMA/CA," *IEEE Journal on Selected Areas in Communications (JSAC)*, vol. 24, no. 7, pp. 1411–1420, 2006.

[3] C. Cano and D. Malone, "On efficiency and validity of previous HomePlug MAC performance analysis," *Computer Networks*, vol. 83, no. C, pp. 118–135, 2015.

[4] V. Ramaiyan, A. Kumar, and E. Altman, "Fixed point analysis of single cell IEEE 802.11 e WLANs: Uniqueness and multistability," *IEEE/ACM Transactions on Networking*, vol. 16, no. 5, pp. 1080–1093, 2008.

[5] G. Bianchi, L. Fratta, and M. Oliveri, "Performance evaluation and enhancement of the CSMA/CA MAC protocol for 802.11 wireless LANs," in *7th IEEE International Symposium on Personal, Indoor and Mobile Radio Communications (PIMRC)*, 1996, pp. 392–396.

[6] *IEEE Standard for Broadband over Power Line Networks: Medium Access Control and Physical Layer Specifications*, IEEE Standard 1901-2010.

[7] OpenWrt. [Online]. Available: openwrt.org/. [Accessed: November 6, 2019].

[8] *Enhancements for Higher Throughput*, IEEE 802.11n-2009-Amendment 5.

[9] M. Heusse, F. Rousseau, G. Berger-Sabbatel, and A. Duda, "Performance anomaly of 802.11 b," in *IEEE INFOCOM International Conference on Computer Communications*, 2003, pp. 836–843.

[10] *IEEE Standard for Information Technology – Telecommunications and Information Exchange between Systems – Local and Metropolitan Area Networks – Specific Requirements Part 11: Wireless LAN Medium Access Control (MAC) and Physical Layer (PHY) Specifications*, IEEE Standard 802.11-1997.

[11] G. Berger-Sabbatel, A. Duda, O. Gaudoin, M. Heusse, and F. Rousseau, "Fairness and its impact on delay in 802.11 networks," in *Proceedings of the IEEE Global Communications Conference (GlobeCom)*, 2004, pp. 2967–2973.

[12] M. Bredel and M. Fidler, "Understanding fairness and its impact on quality of service in IEEE 802.11," in *IEEE INFOCOM International Conference on Computer Communications*, 2009, pp. 1098–1106.

[13] R. K. Jain, D.-M. W. Chiu, and W. R. Hawe, "A quantitative measure of fairness and discrimination for resource allocation in shared computer systems," *Eastern Research Laboratory, Digital Equipment Corporation, Hudson, MA*, 1984.

[14] C. Koksal, H. Kassab, and H. Balakrishnan, "An analysis of short-term fairness in wireless media access protocols," *ACM SIGMETRICS Performance Evaluation Review*, vol. 28, no. 1, pp. 118–119, 2000.

[15] C. Vlachou, J. Herzen, and P. Thiran, "Fairness of MAC protocols: IEEE 1901 vs. 802.11," in *2013 IEEE International Symposium on Power Line Communications and Its Applications (ISPLC)*, pp. 58–63.

[16] S. Klugman, H. Panjer, G. Willmot, and G. Venter, *Loss Models: From Data to Decisions*. John Wiley, 1998.

[17] J.-W. Cho, J.-Y. Le Boudec, and Y. Jiang, "On the asymptotic validity of the decoupling assumption for analyzing 802.11 MAC protocol," *IEEE Transactions on Information Theory*, vol. 58, no. 11, pp. 6879–6893, 2012.

[18] C. Vlachou, "Measuring, modeling and enhancing power-line communications," Ph.D. dissertation, Department of Computer and Communication Sciences, EPFL, Lausanne, Switzerland, 2016.

[19] G. Sharma, A. Ganesh, and P. Key, "Performance Analysis of Contention Based Medium Access Control Protocols," *IEEE Transactions on Information Theory*, vol. 55, no. 4, pp. 1665–1682, 2009.

Part III

Management, Security, and Further Applications

7 Security in PLC

7.1 Introduction

In this chapter, we discuss a major problem in broadcast mediums where every station can hear each of the others, which is maintaining the security and privacy of the transmitted and overheard data. The IEEE 1901 standard and its practical implementations on commodity devices have multiple mechanisms and processes for guaranteeing security and privacy. In Section 7.2, we describe these mechanisms, in particular the passwords, keys, and encryption protocols that PLC devices can use. In Section 7.3, we describe how logical MAC-layer networks are created to maintain data privacy. We also discuss the coexistence of multiple PLC logical networks. In Section 7.4, we present the processes of associating and authenticating a new station by the CCo, and in Section 7.5, we discuss the potential security or privacy issues in current PLC networks. We give some guidelines for guaranteeing the security of a logical network. Finally, Section 7.6 compares Wi-Fi and PLC in terms of security and its robustness, given that both wireless and power line mediums are broadcast in nature and popular in residential networks.

7.2 Security and Privacy Mechanisms

IEEE 1901 and HomePlug ensure security by controlling the access to the network created and configured by the user. In addition, they provide privacy by encrypting all data transferred through the electrical wires and potentially overheard by unauthorized neighboring PLC devices. Access of new stations to the network is provided by using encryption keys and passwords. Encryption relies on 128-AES (advanced encryption standard) and secure hash functions (SHA-256). AES is a widely adopted standard, supported in both hardware and software of commodity networking devices.

7.2.1 Identification, Network Passwords, and Keys

In this section, we describe the keys and passwords used for security and privacy in PLC. Each PLC logical network (as defined later in Section 7.3) has a 54-bit network identifier (NID) that, together with the short NID (SNID), uniquely identifies the network. The NID of a network is chosen by its CCo. A logical network is defined by a set of stations with the same NID and CCo, and as as result, a common network membership key (NMK), which provides a station exclusive data access to the network. NID is broadcasted with every beacon, similarly to the SSID for Wi-Fi, and SNID is included in every frame control.

According to the IEEE 1901 standard [1], the encryption keys used in PLC are 128-bit AES. The standard defines different keys and passwords that are used for different purposes, as described below:

Device access key (DAK) is mandatory and unique per PLC device. It is assigned by the manufacturer and usually printed on the commercial PLC devices along with their MAC address and other information. DAK is used to encrypt messages between two stations, assuming the transmitting station knows the DAK of the destination station. DAK is never transmitted over the power line.

Device password (DPW) is a unique alphanumeric string that can be set by the manufacturer and provided with the hardware. DPW can be printed on commercial PLC devices instead of the DAK. DPW can be used by the user on the vendor's software to associate new stations. After entering DPW, the software will generate the DAK and pass it to the PLC station. We describe how to generate keys from passwords in Section 7.2.2.

Network membership key (NMK) is used by stations to validate their membership in a PLC logical network. A station with valid NMK can join the network. NMK can be generated by the network password using hashing functions.

Network password (NPW) is used to generate NMK using hashing functions and algorithms described in Section 7.2.2. NPW can be provided by the user directly into a new station. This NPW is hashed to create the NMK.

Network encryption key (NEK) is used for encrypting data. Most PLC frames are encrypted by using NEK and the AES protocol. CCo generates the NEK, provides it to its authenticated stations in a message encrypted by the NMK, and does not expose it to the PLC user. NEK may be automatically and dynamically changed, at least once an hour according to the IEEE 1901 standard. CCo has to keep two NEKs, the one being used and a future one.

Temporary encryption key (TEK) is used to encrypt messages between stations in a temporary private channel (also in unauthenticated channels). It can be distributed using the station's DAK and has to be used by at most two stations.

Nonces are pseudo-random numbers that are used only once to prevent replay attacks by adversaries (e.g., data using an existing nonce are invalid). Nonces are particularly useful for MM transmissions.

Encryption key select (EKS) is a 4-bit field in the frame control that is the index of the encryption key used for encrypting each segment. The mapping between EKS and encryption key is provided by the CCo for authenticated stations within the network. EKS value of 0xF is used for unencrypted data.

Other optional keys for 1901 devices can be found in the standard [1].

When a new station attempts to join a PLC network, the authorization procedure determines the compatibility of the user and the network based on the NKM and the station's security level. The security level takes two values depending on whether the NMK is allowed to be exchanged using a unicast key exchange or not. The station can change security level upon the user's configuration, and in a such case, it has to discard the previous NMK. The station authorization by the CCo is attempted if the security levels at which the station and the network operate are the same, if the owner of the new station authorizes participation to the network, and if the network authorizes the station's membership. If these three conditions are satisfied, the station joins the network and obtains the NMK if it does not already have it.

There are three methods to provide NMK to a PLC station: (1) direct entry of NMK (e.g., provided by the user using a software interface) (2) distribution of NMK using the DAK of the PLC stations, and (3) distribution of NMK using a unicast key exchange (UKE) protocol, described in the next paragraph. Method (1) requires the user to know the NMK of the network. Method (2) uses the unique DAK per station, and the user would need to know the DAK of each device. Method (3) assumes that UKE exchange is secure and cannot be decrypted. For instance, it is used when the PLC user employs the "push-button" present on most PLC devices.

After a station has the correct NMK and joins the logical network, CCo provides the current NEK, which is used to encrypt/decrypt physical blocks during segmentation in the MAC layer.

Experimental Guideline The NMK can be directly set to the device by software or manufacturer tools described in Section 4.5.2. The tools `amptool` and `int6k` can set the NMK either for a local or a remote device [2]. To set the NMK on a remote device, the tools request using the remote device's MAC address and DAK.

Unicast key exchange (UKE) is used for communication in unauthenticated channels, between two stations. The final outcome of UKE is a TEK key to provide encryption to subsequent sensitive MM or data. The stations need to send unicast, unencrypted packets to start the UKE process. These messages contain two secret hash keys (one per station), which are then concatenated and hashed to produce the TEK to encrypt the following packets. To minimize the risk of being overheard by adversary stations, these UKE unicast packets should be transmitted with channel-adapted tone maps and not with any broadcast modulation (ROBO modes; see Section 2.5), as we discuss further in Section 7.5. In addition, the messages should not be retransmitted to avoid adversaries having higher probability of retrieving the hash keys. If not received, a new hash key should be generated. After TEK creation, the new key can be used for station authorization or NMK provision.

7.2.2   Key Generations for AES

Key Generation from Password: Keys are generated from a password with the password-based key derivation function 1 (PBKDF1) from PKCS #5 v2.0 cryptography, which is a standardized algorithm from RSA Laboratories public-key cryptography standards (PKCS) series[1]. PBKDF1 from PKCS #5 v2.0 uses a password, a "salt,"[2] and multiple iterations to produce multiple keys from a password. PBKDF1 applies a hash function (SHA-256 for PLC) multiple times to the salt and password, feeding the output of each round to the next one to produce the final output:

$$DAK \text{ or } NMK = PBKDF1(Pwd, S, HF, i, dkLen), \qquad (7.1)$$

where Pwd is the user password, S is a predetermined salt, HF is the SHA-256 function for IEEE 1901, i is the number of iterations, and $dkLen$ is the digest key length ($dkLen = 16$ bytes). The iteration count used to calculate the key is $i = 1000$ for PLC stations. The salt value should be $S = 0x08856DAF7CF58185$ for DAKs and $S = 0x08856DAF7CF58186$ for NMKs, according to IEEE 1901. If the key length is 128 bits, after the thousandth iteration, the leftmost 16 bytes of the SHA-256 output are used as the encryption key (the first byte of the output corresponds to byte 0 of the AES encryption key). The bit ordering of the AES encryption key within a byte is dependent on where it is used.

Passwords are strings chosen from the range 32–127 of ASCII character codes. The length of DPW should be between sixteen and sixty-four characters inclusive. The length of an NPW should be between eight and sixty-four characters inclusive, according to IEEE 1901. Users should be provided a warning by the user interface that security may be compromised if the user enters an NPW fewer than twenty-four characters in length. DPWs should be pseudo-randomly generated and are recommended to be a minimum of twenty characters long.

Automatic key generation for AES is mandatory for PLC devices. All PLC devices should be able to generate hash keys (used in UKE) and AES encryption keys (e.g., NEK, NMK, and TEK) based on random-number generation algorithms. The standard describes a baseline method with minimal security requirements. The vendor implementation should meet the minimum requirements of the baseline.

Nonces generation by devices is also crucial for security. The use of a nonce prevents replay attacks from an unauthenticated station. Nonces are pseudo-random numbers (i.e., the sequence of values that they take for a given station is unpredictable) that are used only once. Practically, as the number of bits in a nonce is finite, they may repeat, but the same nonce should never be used with the same encryption key; that is, changing the encryption key also clears the set of unusable nonces.

📑 **Experimental Guideline** To generate PLC keys from passwords, there are online tools that emulate the SHA-256 hashing. For instance, manufacturer tool `hpavkey` in `Open PLC Utils` (see Section 4.2.4) converts a password to DAK, NMK, or NID [2]. The password should consist of consecutive ASCII characters in the range 0x20 through 0x7F, with a minimum twelve and maximum sixty-four characters. 📑

---

[1] https://tools.ietf.org/html/rfc2898
[2] Salt in cryptography is random data that are used as an additional input to a one-way function that "hashes" data (a password or passphrase). Salts protect against dictionary attacks or their hashed equivalent.

7.2.3   Encryption and Decryption Processes

We provide the main components of the PLC encryption process, as described in the standard. Device-based security network association (DSNA) is the mandatory security algorithm for FFT-based, indoor OFDM PLC systems.[3] DSNA provides authentication mechanisms using the NMK, ensures privacy of data transferred by an NEK, and protects the integrity of data by an integrity check value (ICV). PLC data encryption is performed on the MAC layer as we discussed in Figure 3.1 or within a special management message called CM_ENCRYPTED_PAYLOAD.[4] The PHY-layer approach uses the AES algorithm in cypher block chaining (CBC) mode, which means that the algorithm is performed on fixed-length groups of bits (called blocks and equal to 128 bits for IEEE 1901) and that each block of bits is XORed with the previous encrypted block before being encrypted [3]. All detailed functionalities and definitions for AES can be found in the original proposal [4]. The frame control contains the EKS field to be used for decrypting the PBs. The initialization vector of the AES algorithm is generated from the start-of-frame delimiter (12 bytes), the MPDU PB count (1 byte), and the PHY block header (3 bytes). AES typically runs on multiple iterations (e.g., ten for 128-bit keys) of four processes: The first transformation in the AES encryption cipher is a substitution of 16-byte data using a fixed table; the second transformation shifts data rows, the third mixes columns. The last transformation is an XOR operation performed on each column using a different part of the encryption key.

Figure 7.1 presents the encryption process for IEEE 1901. The MSDUs are assembled into the MAC frame stream as also described in Section 3.2. Each MAC frame is appended with an integrity check value (ICV). The MAC frame stream is segmented into PBs (typically 512 bytes for data), and the EKS of the key is inserted into the MPDU header. PBs are encrypted individually. AES CBC requires a unique nonce value for each frame protected by an encryption key, and DSNA CBC derives a 128-bit initialization vector (IV) for this purpose. Each PB is given a header and is encrypted using the NEK and the IV constructed from the MPDU header, the PB header, and the location of the PB in the MPDU, as we see in the figure. The PB is then given a check sequence (PBCS) computed using the PB header and the encrypted segment, which forms the PB body. The three elements – PBCS, body (encrypted data), and header – form the encrypted PHY block, which is then sent in the MPDU. If the segment has to be re-sent, it is encrypted again for each transmission, as the key can change, and the IV is likely to have changed, since it depends on the PB header and location in the MPDU. Figure 7.2 shows the corresponding process for the decryption of a MAC frame stream.

7.2.4   User Interface Station (UIS)

A network administrator is able to control the creation of a PLC logical network, modify the default passwords for security, and assign different privacy domains (logical

---

[3] The standard also introduces optional robust security network association (RSNA) algorithms that are outside the scope of this book. In addition, wavelet PLC uses pairwise-based security network association.

[4] The difference with this method compared to the MAC-layer approach is that only part of the payload is encrypted.

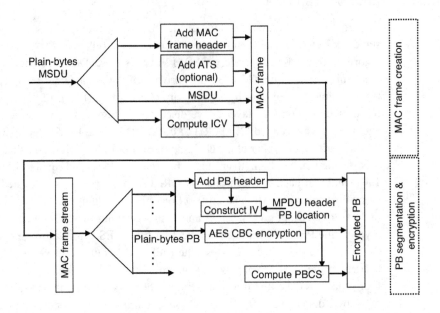

Figure 7.1 IEEE 1901 MAC frame generation and DSNA encryption processes. IEEE 1901-2010. Adapted and reprinted with permission from IEEE. © IEEE 2010. All rights reserved.

networks). This can be done through a user interface station (UIS), which is a station that enables interactions with the PLC network, security, and control configurations. Most PLC stations have UIS capabilities, as we observe in the market (vendors such as Netgear, D-Link, and Devolo provide software UIS to configure the network). In Chapter 4, we provide multiple examples of configuration using UIS and management messages. Examples of UIS software are shown in Figures 4.4 and 4.9. The communication between the UIS and the PLC device is done without encryption and through Ethernet on the MAC layer. UIS might not be physically connected to the PLC network. For example, in Figure 4.1, we see that it can connect through a bridging station or interface.

7.3   PLC Logical Networks

A logical network is defined as a set of stations that share the same cryptographic key (AES/NMK) and that have the same CCo over the power line medium. We describe the creation of a PLC network and the coexistence between multiple networks over the same power line medium. When a station discovers an existing logical network (LN) with the same NMK and security level (SL), it joins this network. To join a logical network, a station must have the following:

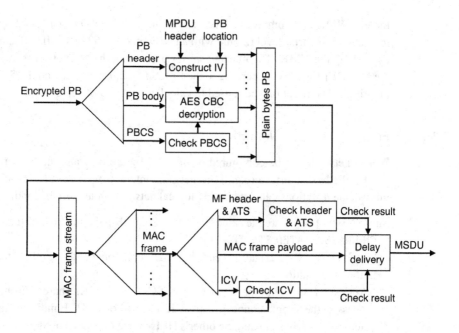

Figure 7.2 IEEE 1901 DSNA decryption process. IEEE 1901-2010. Adapted and reprinted with permission from IEEE. © IEEE 2010. All rights reserved.

- A valid NMK and security level, compatible with the network. NMK is obtained by the processes described in Section 7.2.1.
- A unique TEI, which is obtained from the association with the CCo, a procedure described later in Section 7.4.1.
- The current NEK, which is obtained by authentication, a process presented in Section 7.4.3.

There are two security levels defined in the standard: (1) Simple connect (SC), where NMK may be exchanged using unicast key exchange, and (2) Secure (HS), where NMK should not be exchanged using UKE, that is, the NMK must be configured manually. If two or more stations with the same NMK and SL can hear each other but there is no LN, the stations will form a new LN. The simple connect function can be implemented with a push button on the device or using the user interface station.

📲 **Experimental Guideline** Most commodity PLC devices support simple connect functions, using a pairing "push-button" on the device. To automate such a procedure, the amptool described in Chapter 4 provides commands to use the simple connect function in different modes (e.g., join or leave network, status, and reset) [2]. 📲

A logical network has a unique 54-bit network identifier (NID), which changes only when the NMK changes. Upon NMK modification, every authenticated station receives the new NMK and computes or receives the new NID. NID is generated by combining the network security level (2 bits) with an NID offset of 52 bits. NID offset is generated by hashing the NMK. The mechanism for generating the NID offset from the NMK is the PBKDF1 function (according to the PKCS #5 v2.0 standard) and the SHA-256 hash algorithm. The iteration count used to calculate the NID offset is 5 and no salt is used.

7.3.1   Forming a Logical Network

A new network can be configured or initiated by the user and the higher-layer entity (HLE), the host interface (e.g., Ethernet) that interacts with the PLC MAC layer. Two unassociated stations can form a new logical network under certain conditions:

Same NID: They share the same NID and SL, NMK pair, and both of them receive the CM_UNASSOCIATED_STA.IND MM.

DAK-Encrypted: One station sends its NMK, encrypted with the other station's DAK, to the other station.

SC-Add/SC-Join: They both have NMK-SCs (i.e., NMKs shared through the UKE process in the simple connect security level), and one's HLE indicates that it should enter the SC-Join state and the other's HLE indicates that it should enter the SC-Add state.

Both in SC-Join: They both have NMK-SCs, and the HLE of each indicates they should enter the SC-Join state.

To form the network, the stations exchange MM with their CCo capabilities. One of them will become the CCo, and the other one will associate and authenticate with the new CCo. We describe how the stations form the network using the four methods above.

Same NID: The stations transmit the CM_UNASSOCIATED_STA.IND MM, which contains the NID and the CCo capability of the station. If a second station receives this message and has a matching NID, it decides whether it or the other station becomes a CCo. A matching NID and security level means that the stations will also have the same NMK (configured by the user or the manufacturer). If the station decides to become the new CCo, it starts sending beacons with their NID, and the other stations try to associate with the new CCO by sending the CC_ASSOCIATION.REQ MM.

DAK-Encrypted: In this case, the HLE (or UIS) provides a station with the DAK of another station aiming to pair the stations and form a new network. The station with the DAK forms a new CM_SET_KEY.REQ MM that contains a TEK encrypted with the target station's DAK. This MM is broadcast, and any overhearing station attempts to decrypt the MM. Only the station with the specific DAK can decrypt the message, and receive the TEK. In such a case, it replies with the CM_SET_KEY.CNF MM, which contains the NMK of the network. This MM is encrypted with the TEK. Then, one of the stations becomes the CCo, and the other station associates and authenticates with the CCo.

SC-Add/SC-Join: A station in SC-Join state transmits a CM_SC_JOIN.REQ MM periodically until it joins a network or there is a timeout. A station in SC-Add waits to receive this MM until it either adds this station to the network or there is a timeout. Upon the reception of the MM, the station in SC-Add state responds with the

CM_SC_JOIN.CNF MM, becomes the CCo of the network, and starts transmitting beacons. The station in SC-Join state begins its association and authentication procedure upon receiving the first beacon, regardless of its CCo capabilities. PLC channels are highly frequency selective among different links. After association, there is an optional sounding and tone map estimation process before exchanging any keys over the power line. This enhances physical security as an adversary would need to capture all tone map frames and to have a similar or better channel than the receiver of the message to be able to demodulate and decode the frame (see Section 7.5).

Both in SC-Join: If PLC stations are set in join state (e.g., through the HLE or the push-button on the device), then they both send a CM_SC_JOIN.REQ MM with their CCo capabilities using unencrypted, multinetwork broadcast. Upon the reception of such an MM, the station decides whether to be the CCo. If the station decides not to be the CCo, it sends a CM_SC_JOIN.CNF MM and starts waiting for the beacons of the CCo upon which it tries to associate. After association, there is an optional sounding and tone map estimation process before exchanging any keys over the power line, as in the previous method of forming a network.

The detailed formats of all aforementioned MMs can be found in the IEEE 1901 standard [1].

7.3.2   Power-On Network Discovery Procedure

In the previous section, we have described the methods to form a logical network. We now discuss the sequence of events when a PLC station is powered and tries to form a logical network or join one. After the power-on procedure, the station can have one of the following states: (1) unassociated station, (2) unassociated CCo, (3) associated station, or (4) CCo.

A CCo-capable station runs the power-on network discovery procedure to detect whether another PLC network is active or if a new one can be formed. Before starting the power-on network discovery procedure, the station selects a Beacon Backoff Time (BBT). BBT is the maximum duration of time for which a station performs the power-on network procedure. If the station was the CCo before it was powered down, BBT should be chosen as a random value in the interval (MinCCoScanTime=1 s, MaxCCoScanTime=2 s). Otherwise, BBT should be a random value in the interval (MinScanTime=2 s, MaxScanTime=4 s). Upon the expiration of the BBT, the station processes all the received unassociated station information (if any). If the station detects other unassociated stations with the same NID, it should decide whether to become a new CCo and form a new network. Otherwise, it should become an unassociated CCo. A station that fails to form or join a network during the power-on procedure operates in unassociated station mode or unassociated CCo mode.

During the power-on network discovery procedure, if one or more networks are detected, the station should attempt to transmit an CM_UNASSOCIATED_STA.IND MM using multinetwork broadcast approximately once per unassociated station advertisement interval (USAI, 1 s) to provide information to other stations that may be performing the same procedure. The transmission time of each CM_UNASSOCIATED_STA.IND is randomly chosen. An unassociated station sends the `CM_UNASSOCIATED_STA.IND` MM periodically once per MaxDiscoveredPeriod (10 s). It continues to operate in this mode unless it detects a network with matching NID or it determines that there are no neighboring networks and becomes an unassociated CCo. If the station detects an existing network or station with matching NIDs, it follows the procedures described in Sections 7.3.1 and 7.4.1. A station can enter the unassociated station mode after power-on discovery, after its logical network no longer exists, or after being asked to leave the logical network.

An unassociated CCo keeps monitoring the medium for other networks or stations attempting to join its network. It can become an unassociated station, if it detects another PLC network or a CCo, or if it receives an association request from another station. A station can enter the unassociated CCo mode after power-on discovery or after all other stations have left the logical network. Recall that every PLC station needs to synchronize each clock based on the detected beacons that are in line with the AC power cycle. The station synchronizes its PHY clock (PHYClk) to one of the detected CCos and should use the SNID of that CCo in multinetwork broadcast transmissions. This provides a way for other stations that also hear the same CCo to apply the appropriate PHYClk correction for reception.

The power-on procedure starts when the station is powered on and also when the station cannot detect the beacon of its existing logical network. The detailed process is summarized in Figure 7.3.

Multinetwork broadcast (MNBC): In the power-on discovery procedure, some management messages require reception by all stations, including the ones not associated with any CCo and stations associated with other CCos. Such MM transmissions use the MNBC transmission mechanism. Each MNBC transmission is unencrypted and includes an RTS/CTS exchange (see Section 3.6.2 for definition of these control frames) before the MM is transmitted. RTS/CTS notifies the receiving stations to update the virtual carrier sense clocks (Section 3.5.2) for the long MPDU that follows. CTS can be transmitted by a broadcast/multicast proxy, but it is not required. To enable MNBC, the transmitting station should set the multinetwork broadcast flag (MNBF) in the start-of-frame delimiter, in the RTS, and in the CTS. The flag indicates that the MPDU is a broadcast message to all stations regardless of the SNID or network association. All stations attempt to decode any MPDU when MNBF is set to 0b1.

7.3.3   Coexistence of Multiple Logical Networks

Similarly to Wi-Fi logical networks, which can coexist over the same medium over the air, PLC logical networks can coexist due to the broadcast nature of the power line medium. In an enterprise environment, the network administrator might create different PLC networks for security and privacy purposes. Although the networks would not communicate with each other, except for the CCos, as shown in Figure 7.4, they could potentially share the physical medium and interfere (i.e., share the airtime and throughput). In a residential environment, it is likely that every household has a different PLC

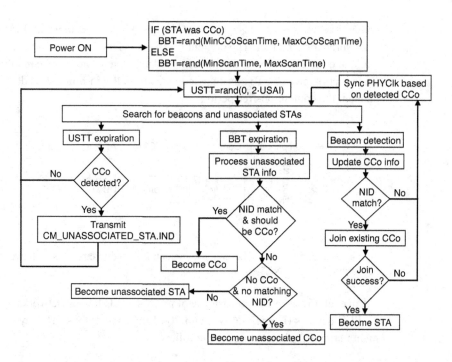

Figure 7.3 The power-on discovery process of IEEE 1901. IEEE 1901-2010. Adapted and reprinted with permission from IEEE. © IEEE 2010. All rights reserved.

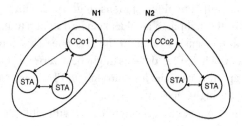

Figure 7.4 Coexistence of two logical networks, N1 and N2.

logical network, managed by each residence. There, too, networks could interfere due to cable crosstalk.

When multiple PLC logical networks are present, each network has a unique NMK and a CCo that broadcasts beacons and manages its network. The stations of one logical network are able to communicate between each other and decrypt data on the MAC layer. The beacons of all logical networks are broadcasted over a specific period after the AC power phase crosses 0 value (AC power is a sinusoidal wave). The offset after zero-crossing on which beacon transmission is expected is included in the beacon. Figure 7.5 shows the transmission periods for multiple logical networks (e.g., networks "N1" and "N2"). The beacon communicates the scheduling and allocations for CSMA/CA and contention-free sessions. The CSMA/CA region is shared among all PLC networks that can overhear each other. The contention-free slots are strictly assigned to a single network, and the exact slots are communicated and determined by the involved CCos after a negotiation process. To our knowledge, commodity in-home PLC devices do not implement the contention-free period.

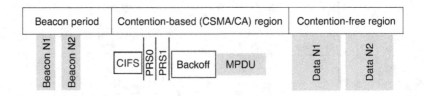

Figure 7.5 Example of PLC beacon and contention-based/free periods with two logical networks "N1" and "N2."

Liu et al. [5] study the coexistence of multiple PLC logical networks, using commodity PLC devices (HomePlug AV with 200 Mbps and 1 Gbps PHY-layer capacity). The main findings of the authors are the following.

- PLC networks can communicate when they are set up on different phases and even on separate unsynchronized wires. Hence the shared medium for PLC can cover a large area in a residential/enterprise building during crosstalk between wires.
- Newer standard HomePlug stations (e.g., AV2 with 1 Gbps capacity) are forced to work at low throughput, if an older standard station (e.g., AV with 200 Mbps capacity) operates at the sensing range of the first one. This could have impact in residential deployments where PLC devices might be diverse (users might buy PLC devices from different apartments). The low-throughput PLC devices can bring down the capacity of up-to-date high-capacity devices.
- Per-link throughput outages might occur with multiple logical networks on different power line phases. The authors discovered this issue experimentally, but the cause is unknown. Hence it might be due to the chipset manufacturer.
- Sudden loss of CCo can lead to disconnection and outage of up to 10 s. This might mean that the PLC network has not provisioned a backup CCo, and it can be due to vendor-specific implementations.

7.4   Station Authentication Procedure

7.4.1   Station Association

Station association is the process of obtaining a valid TEI from the CCo. The station may try association with the CCo of a network. There is a single method for association and multiple methods for deassociation. Upon deassociation, the TEI becomes invalid until it can be reassigned to another station by the CCo. TEI is 8 bits long and its value for the station is 0x00 if not yet assigned and is in the range 0x01–0xFE when associated. The TEI value of 0xFF is only used for broadcast transmissions. When a new station is associated, the CCo provides the complete TEI map (all other associated stations) to that station and updates all other stations in the logical network with the new TEI map information. CCo provides a lease time for every new TEI, which can be from 15 min to 48 hours. If the current lease expires and the CCo has not renewed it, the station has to stop using its TEI and apply for another TEI. The association process and the MMs used during it are depicted in Figure 7.6. The exact content of the MMs can be found in the IEEE 1901 standard [1].

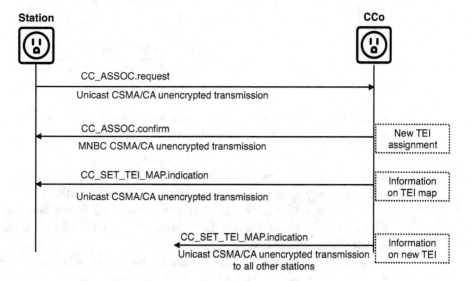

Figure 7.6 Association process for a new station and the MMs used. IEEE 1901-2010. Adapted and reprinted with permission from IEEE. © IEEE 2010. All rights reserved.

7.4.2 Station Deassociation

Station deassociation occurs when the PLC device is powered off or when the user prompts the device to leave the logical network. The station has to notify the CCo that it is about to leave the network and wait for a response before actually leaving. Similarly to the association process, the CCo informs the rest of the stations about the updated TEI map. The deassociation process and the MMs used during it are depicted in Figure 7.7.

7.4.3 Station Authentication

To authenticate with a CCo, the station needs a valid NMK, which can be obtained with three methods described in Section 7.2.1. After the station is associated and has obtained a valid NMK, it uses the NMK and its TEI to join the PLC logical network.

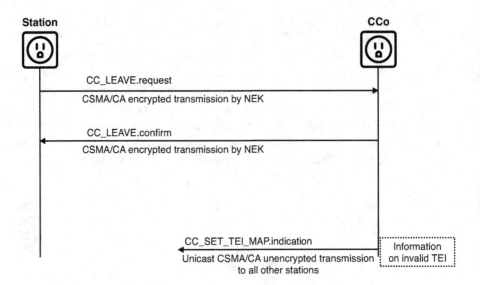

Figure 7.7 Deassociation process and the MMs used. IEEE 1901-2010. Adapted and reprinted with permission from IEEE. © IEEE 2010. All rights reserved.

If the station's NMK is validated by the CCo, CCo provides an NEK, which is the authentication procedure. If the CCo cannot validate the NMK, then the station flags this key as invalid and can either try again to obtain a valid NMK or connect to another logical network. CCo maintains the authentication status for the station as long as the association status is also valid.

Figure 7.8 shows the IEEE 1901 authentication process. The procedure starts with the station sending the MM CM_GET_KEY.request, containing the key type under request, that is, the NEK, the station's TEI from the association, and its MAC address. This MM also contains a newly generated nonce, and it is encrypted within a CM_ENCRYPTED_PAYLOAD.indication MM, by using the NMK. When the CCo validates that the correct NMK was used to encrypt the message, it replies with the NEK and EKS to the station. Similarly to the request, the CCo encrypts the NEK and EKS MM (called CM_GET_KEY.confirm) into a CM_ENCRYPTED_PAYLOAD.indication MM. The NEK can be used immediately after reception, for data encryption and decryption. In case of failure to decode the CM_GET_KEY.confirm MM, the CCo replies with CM_ENCRYPTED_PAYLOAD.indication and a "failure" result. The station can flag the NMK as invalid and try to reobtain an NMK for the CCo.

7.5   Eavesdropping PLC Data

In this section, we discuss potential security threats for PLC security and privacy mechanisms. We first discuss how an adversary could capture, demodulate, and decode PLC signals. Then, assuming that the adversary can capture these signals, we discuss

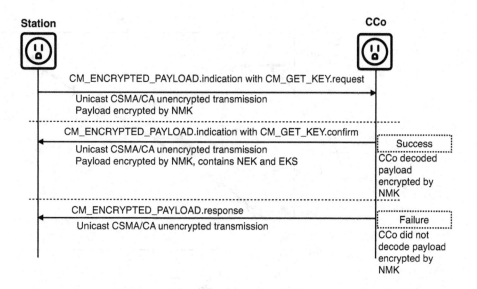

Station | CCo

CM_ENCRYPTED_PAYLOAD.indication with CM_GET_KEY.request

Unicast CSMA/CA unencrypted transmission
Payload encrypted by NMK

CM_ENCRYPTED_PAYLOAD.indication with CM_GET_KEY.confirm

Success
CCo decoded
payload
encrypted by
NMK

Unicast CSMA/CA unencrypted transmission
Payload encrypted by NMK, contains NEK and EKS

CM_ENCRYPTED_PAYLOAD.response
Unicast CSMA/CA unencrypted transmission

Failure
CCo did not
decode payload
encrypted by
NMK

Figure 7.8 Authentication process and the MM used. IEEE 1901-2010. Adapted and reprinted with permission from IEEE. © IEEE 2010. All rights reserved.

potential traffic attacks on these signals' data. To capture PLC data, an adversary must be able to (1) demodulate and decode the signals, i.e., transform the signal into data bits, and (2) decrypt them, that is, transform the raw bits into originally transmitted information. We study these two aspects separately.

7.5.1 Demodulating and Decoding PLC Signals

As described in Section 2.5, each OFDM subcarrier uses a different modulation to encode the information. To be able to decode the signal, a PLC device must have the knowledge of the entire tone map (see Section 2.7). Demodulating the signal requires quite some information, such as the modulation scheme for all the 917 carriers in HomePlug AV, and then using the FEC and interleaving functions to decode the bits. Hence, an adversary would first need to operate in monitor mode and capture the sounding packets during channel estimation and the up-to-date tone maps.

Commodity PLC devices are not able to function in monitor mode (i.e., to decode messages exchanged between two other stations). In addition, since the firmware of the PLC devices is typically proprietary, it would prove very challenging to modify an existing PLC device to work as a monitor interface. To have a chance to decode a PLC signal, one must consequently be equipped with more complex equipment, such as a software-defined radio [6]. SDR is very flexible and can be used to map the signal to bits. With PLC, broadcast messages and network-level MMs are sent with the most robust modulation (ROBO modes), which means that anyone on the same power line network – for example, a neighbor in a multiapartment building who might often be able to see the most robust modulation – can, with the right equipment, decode these

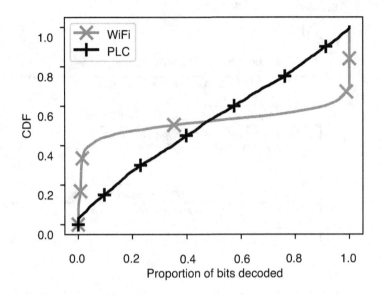

Figure 7.9 Proportion of decodable bits for Wi-Fi and PLC, considering an eavesdropper in the network.

messages. In particular, they can access the knowledge of the tone maps for all links of the network, because tone maps are sent in ROBO mode.

However, knowing the tone map is not enough to be able to decode the signal: the attenuation on the link between the source and the eavesdropper will likely be different from the attenuation between the source and the destination (i.e., on the link on which the transmission occurs). Even if the eavesdropper knows which modulations are used for which links, she must receive a signal strong enough to be able to demodulate and decode it: if the source and the destination are very close, they will be able to use high modulations, which an eavesdropper farther away will not be able to demodulate even if she knows which modulation is used, as the signal will be much more attenuated for her and the SNR will be lower.

The fact that PLC has a different modulation scheme per subcarrier plays an important role: the proportion of the signal that an eavesdropper will be able to decode will be very different compared to Wi-Fi. Wi-Fi uses a single modulation for all the subcarriers: if the eavesdropper is close enough, she will be able to decode virtually the entirety of the signal; if she is too far, she won't be able to decode anything. With PLC, this is different. Because of the high complexity of the noise patterns, with complicated multipath effects, even an eavesdropper far away from the source might be able to decode some carriers, and a station much closer to the source than the destination might be unable to decode some subcarriers. This is illustrated in Figure 7.9, where we show the proportion of the bits that an eavesdropper could theoretically decode, when the link and the eavesdropper are randomly chosen on a testbed of twenty stations (see the testbed in Figure 5.1). This is computed as follows: we gather the tone maps for the two links, source destination (S–D) and source eavesdropper (S–E). For each subcarrier, if the modulation scheme (e.g., 8/16/64/256/1024/4096 QAM) for the link S–E is at least as high as the modulation scheme for the link S–D, we assume that the eavesdropper is able to demodulate the signal. Otherwise, she is not able to demodulate it. We use this to compute the proportion of the number of bits that an eavesdropper would be able to decode. For Wi-Fi, we show the number of bits that have been correctly decoded by the eavesdropper through a monitor interface.

We notice that the shapes of the curves are very different for Wi-Fi and PLC. As can be expected, Wi-Fi is close to being binary: either the eavesdropper is close to the stations, and she can decode all the bits, or she is far away, and she cannot decode anything. With PLC (which uses modulations per subcarrier with complex noise patterns), the shape of the curve is very close to a linear function. With PLC, forward error correction codes (described in Section 2.4) are used in most cases with a rate 16/21. As a first approximation, and ignoring the channel interleaver, we assume that if an eavesdropper is able to decode at least 16/21 of the bits, she is able to decode the entire message. With this assumption, in our experimental setup, about 47 percent of the Wi-Fi messages can be decoded, whereas this is the case for only 25 percent of the PLC messages. Qualitatively, PLC appears more complex to decode than Wi-Fi.

7.5.2 Decrypting PLC Data

If an eavesdropper succeeds in demodulating and decoding a PLC signal, she still has to decrypt it with the AES algorithm to get a plain-text message. Decrypting the message requires knowing the NEK, as explained in Section 7.2.1, which means (1) knowing the NMK, and (2) authenticating with the CCo.

7.5.2.1 Knowing the NMK

In a well-configured network, an eavesdropper has no easy way to retrieve the NMK. Ideally, as far as security is concerned, the NMK should be generated from a pass phrase that is hard to brute-force, and each time a new device is added, this device should be configured with this NMK through the vendor's software or a command line interface. This is also the goal of the HomePlug/IEEE "secure" mode, which requires manual key configuration (see Section 7.3). However, one important aspect of commercial PLC is its "plug-n-play" nature and easy installation. For this reason, many vendors design their device in "simple connect" default mode and use the same default NMK across devices, corresponding to the network password "HomePlug." This was observed before [7], and we have verified experimentally that it is the case for the vast majority of the vendors. If the user does not change this password or NMK, any eavesdropper would know the NMK and can generate it from password "HomePlug." The eavesdropper can then authenticate with the CCo of the logical network. It can also become the CCo and manipulate all network configurations.

Most devices also have a "secure push-button." When this button is pushed, the NMK is changed and sent encrypted with the TEK that is derived from messages sent in clear between stations, which requires unicast messages in both directions. This means that an eavesdropper that arrives in the network later than the transmission of these unicast messages cannot retrieve the NMK. An eavesdropper who was present when the secure button was pushed may know the NMK only if she was able to decode the unicast messages exchanged between the stations [8] – which, as we have shown earlier, is not straightforward given the tone map modulations.

7.5.2.2 Setting the NMK

Commodity tools enable setting the NMK of a remote PLC device on the power line, as long as the DAK of the device is known. Prior studies have shown that DAK can be predicted given the PLC device's MAC address [7]. The authors find that the device password is often produced from the MAC address, and DAK is computed from this password. Many vendors can be affected by a DAK attack when it depends on the MAC address. The MAC address of a PLC device can be retrieved by sniffing PLC frames with faifa [9]. For example, the MAC address of the CCo of another network can be retrieved by capturing the beacons and looking for byte sequences with known vendor OUIs (e.g., Qualcomm/Atheros chipsets would start with bytes 00:50:C2). The eavesdropper can change the NMK of the CCo simply by capturing the beacons and generating the predictable DAK of some vendors. She could also change the NMK of all other stations after this operation, so that the NMK change is not detected over the entire network. Dudek has developed open source software tools to emulate such attacks [10].

📇 **Experimental Guideline**   To re-create the DAK from the MAC address for some PLC vendors, the intruder has to run the following commands from Open PLC utils [2]:

```
1      pass=$(mac2pw -q <MAC_ADDRESS>)
2      DAK=$(hpavkey -D $pass)
```

7.5.2.3 Authenticating with the CCo

The NMK is not used directly for encrypting the PLC messages. Instead, a message is encrypted with the NEK. The NEK is based on AES-128, which is broadly considered to be a secure protocol. Any station that knows the NMK can authenticate with the CCo and obtain the NEK. Guessing directly the NEK by an intruder is not viable, as it can be modified by the CCo at any time (period not defined by the standard), and brute-forcing can typically take longer than the validity period of the NEK. In theory, an eavesdropper with a valid NMK could also become the CCo, retrieve the network structure and the MAC addresses of the stations, and modify the NMK, taking control of the logical network.

7.5.2.4 PLC Metadata

PLC emits electromagnetic radiation. It is possible to detect this radiation to discover the presence of a network, even without knowing the NMK, when located a few meters away from a device [11]. Observing PLC metadata or the presence of a PLC network is usually much easier to do – but also gives much less information unless it contains information on the MAC address for DAK attacks descibed above. The eavesdropper can retrieve beacons, and any delimiter that contains information about setting the network allocation vector and deferring to transmissions (see Section 3.5.2). An eavesdropper can get all metadata and broadcast messages at the network level (i.e., management messages), and it can detect if a message is sent between two stations (without being able to decode the message itself). This can be achieved with any commodity device, by using standard open source tools such as faifa [9].

7.5.3   Summary

If well configured, PLC is a secure technology, as it can only operate in encrypted mode – that is, all PLC messages will be encrypted and hard to decrypt. However, to make it easy to use, PLC vendors often use a default security key, which makes it easier for an eavesdropper to associate with the CCo and decrypt any data. It is hard for this association to go unnoticed, because the eavesdropper would still need to authenticate with the CCo, or become the CCo of the network to obtain the NEK for data decryption. In addition to vulnerabilities due to the default NMK, the DAK of the devices can be predictable with some vendors, allowing an intruder to modify the device's configurations and NMK. In addition to data eavesdropping, an intruder knowing the NMK can destroy the PLC devices by uploading certain firmware versions that enable erasure of the NVRAM [7].

 Experimental Guideline   To detect an intruder or changes in logical network topology, the network administrator can run the following command from Open PLC utils [2] and monitor the authenticated stations, MAC addresses, and the CCo of the network:

```
1        ampstat -m -i <ETH_INTERFACE> # Read network membership
             information
```

Eavesdropping PLC data is challenging due to the increased complexity required to decode PLC signals. Overall, eavesdropping unicast PLC messages modulated with non-ROBO tone maps, though not impossible in many cases (in simple connect mode), is an intricate task that requires complex hardware and is not ensured to succeed. To fully secure the network on the PHY layer, management frames containing tone maps should not be transmitted unencrypted. In addition, NMK, and other security information should be modulated via non-ROBO tone maps between the station and the CCo. Getting metadata about the PLC frames is easier compared to eavesdropping actual data, as start-of-frame delimiters are transmitted at ROBO modulations and demodulated by any PLC device; some of these metadata attacks are easy to prevent by re-configuring the network, but some of them stem from the very nature of PLC communications. However, metadata attacks give little information, which would be the presence or absence of a network or of a communication link between two stations. Metadata attacks can be harmful when PLC devices have poorly configured or predictable DAK. Table 7.1 summarizes guidelines on guaranteeing secure PLC logical networks for vendors and users.

Table 7.1 Guidelines for PLC security

Guideline	Purpose
Monitoring network with `ampstat` periodically	Detect MAC address of intruder(s)
Configure a strong network password	Prevent brute-force NMK prediction
Change NMK periodically	Avoid NMK attacks
Configure random DAK (vendor)	Avoid DAK reverse engineering
Unicast all MMs with security information and tone maps (vendor)	Rely on PHY-layer security and tone maps to avoid decoding signals

7.6   Comparison with Wi-Fi

We now summarize the differences between Wi-Fi and PLC in terms of security.

7.6.1   Security Protocols on MAC layer

HomePlug AV devices must use 128-bit AES-based encryption in CBC mode, which is known to be cryptographically superior to non-CBC encryption and to offer strong security guarantees. In contrast, the IEEE 802.11 standard does not enforce any encryption, and some Wi-Fi networks still use an open network. Open networks typically operate in public areas, such as airports, educational institutions, restaurants, and cafeterias. In open networks, the user should ensure that application layer security is applied on their data transmission (e.g., through hypertext transfer protocol secure [HTTPS]). Wi-Fi networks can be used with security protocols. Typically, residential and enterprise networks are private and use security protocols. The first of these security protocols was WEP (Wired Equivalent Privacy); it is based on another encryption scheme (RC4) and does not use CBC, and it is considered as easy to break. It has been replaced by WPA (Wi-Fi Protected Access) protocols, in particular by WPA2, which is currently the most prevalent security protocol for Wi-Fi. WPA2 is also based on 128-bit AES and uses a CBC message authentication code (CBC-MAC). With this protocol, the security guarantees for PLC and Wi-Fi are similar.

Note that even though the PLC standard enforces a strong security protocol, we have shown in Section 7.5 that many out-of-the-box devices use a default key from which the encryption key is derived, which makes it easy to break if ill configured.

In PLC, two modes of operation are possible: simple connect, in which the key is exchanged over the medium, and secure mode, in which the key needs to be manually configured. In most home networks, the simple connect mode is used; in contrast, Wi-Fi keys are usually manually configured by entering a pre-shared key. The simple connect mode is similar to the Wi-Fi protected setup (WPS) mode, which is an optional standard for easy and secure establishment of a Wi-Fi network either with a push-button or with a personal identification number.

7.6.2 PHY Layer Robustness

As we have seen in Section 7.5, the differences between PLC and Wi-Fi at the PHY layer make it much more difficult to decode PLC signals. With Wi-Fi, the unique MCS is included in the preamble, and thus an intruder knows directly how to demodulate and decode the packet; in PLC, there is one modulation scheme per carrier, and it is more difficult to obtain. This also means that decoding PLC signals requires more complex equipment, such as software-defined radios, as opposed to Wi-Fi, for which a commodity device is typically good enough. In addition, Wi-Fi offers mobility, and an intruder can approach the source of the signal more easily. In contrast, a PLC intruder has to be physically plugged into the same or an adjacent PLC network and to have at least the same channel quality as between source and destination of the message to guarantee a low error rate in demodulating the signal.

7.7 Summary

In this chapter, we have described the security mechanisms used in PLC. PLC devices are required to use a 128-bit AES-based security protocol, known to offer strong security guarantees. We have presented the procedures for forming a logical PLC network and for incoming devices to associate and authenticate with this network. We have presented the different security modes, simple connect in which the key is exchanged over the medium (lowering the security guarantees but easing the authentication), and secure mode, in which the keys are manually configured for each device.

We have shown that even though some security mechanisms might be easily breakable, in particular in ill-configured networks, the PHY layer channel estimation of PLC makes it typically extremely difficult for an eavesdropper to decode PLC signals between two other devices. This also means that a well-configured PLC network can offer strong PHY-layer security guarantees.

References

[1] *IEEE Standard for Broadband over Power Line Networks: Medium Access Control and Physical Layer Specifications*, IEEE Standard 1901-2010.

[2] Atheros Open Powerline Toolkit. [Online]. Available: github.com/qca/open-plc-utils. [Accessed: November 6, 2019].

[3] W. F. Ehrsam, C. H. Meyer, J. L. Smith, and W. L. Tuchman, "Message verification and transmission error detection by block chaining," US Patent 4 074 066, Feb. 14, 1978.

[4] J. Daemen and V. Rijmen. (1999) AES proposal: Rijndael (version 2). [Online]. Available: www.cs.miami.edu/home/burt/learning/Csc688.012/rijndael/rijndael_doc_V2.pdf. [Accessed: November 6, 2019].

[5] Z. Liu, A. El Fawal, and J.-Y. Le Boudec, "Coexistence of multiple HomePlug AV logical networks: A measurement based study," in *Proceedings of the IEEE Global Communications Conference (GlobeCom)*, 2011, pp. 1–5.

[6] N. Otterbach, C. Kaiser, V. Stoica, B. Han, and K. Dostert, "Software-defined radio for power line communication research and development," in *Workshop on Software Radio Implementation Forum*, 2015, pp. 37–42.

[7] S. Dudek. (2015) HomePlugAV PLC: Practical attacks and backdooring. [Online]. Available: www.synacktiv.com/ressources/NSC2014-HomePlugAV_attacks-Sebastien_Dudek.pdf. [Accessed: November 6, 2019].

[8] R. Newman, S. Gavette, L. Yonge, and R. Anderson, "Protecting domestic power-line communications," in *ACM Symposium on Usable Privacy and Security (SOUPS)*, 2006, pp. 122–132.

[9] X. Carcelle and F. Fainelli, "FAIFA: A first open source PLC tool," 2008, [Online]. Available: fahrplan.events.ccc.de/congress/2008/Fahrplan/events/2901.en.html. [Accessed: November 6, 2019].

[10] HomePlugPWN. [Online]. Available: github.com/FlUxIuS/HomePlugPWN. [Accessed: November 6, 2019].

[11] R. Baker and I. Martinovic, "EMPower: Detecting malicious power line networks from EM emissions," in *33rd IFIP International Conference on Information Security and Privacy Protection*, 2018, pp. 108–121.

8 Heterogeneous Networks and IEEE 1905

8.1   Introduction

To address the always-increasing demand for high performance in residential or enterprise networks, significant effort has been made to improve the standards for the different technologies used in these networks: HomePlug AV2 for PLC; IEEE 802.11ac/ax for Wi-Fi; 10 Gigabit Ethernet; MoCA 2.5 for coaxial communications. But much higher performance can also be achieved by using these technologies simultaneously, forming so-called heterogeneous (or hybrid) networks.

Because the technologies of heterogeneous networks do not interfere with each other, high multiplexing gains are possible. In addition, when the underlying physical mediums are different (e.g., Wi-Fi and PLC or coaxial cables), the potential reasons for weak signals are different, and this increased spatial diversity enables a better coverage and a reduction of the weak spots in local networks [1].

Using different technologies in residential and enterprise networks is not new. But most often, each technology is used for a pre-determined role: Wi-Fi is used as a means of providing easy access to mobile clients, and wired links serve as a backbone and provide access to fixed clients. For example, in enterprise networks, Wi-Fi access points are typically connected together with Ethernet cables, or, in home networks, Wi-Fi is used for mobile access and PLC, Ethernet, or MoCA for fixed devices (desktop computers, televisions, etc.).

The IEEE 1905.1 standard aims at making interoperability between the technologies easier and more flexible. By introducing an abstraction layer between the MAC and IP layers (layer 2.5) and a common interface, it enables a device to seamlessly use or alternate between two or more technologies, depending on the network conditions. Its goal is also to make the management of heterogeneous networks simpler, by providing end-to-end QoS, autoconfiguration, and common secured authentication and setup methods.

In this chapter, we present different standardized solutions for heterogeneous networks. We describe in particular the IEEE 1905.1 standard. We also present algorithms that can be used in heterogeneous networks to improve performance.

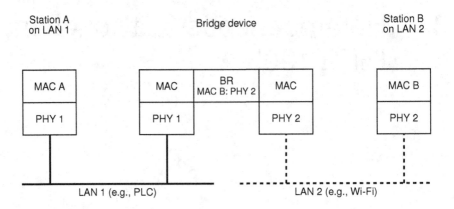

Figure 8.1 A network bridge on a single IP network. The mapping between the MAC address and the outgoing interface is shown.

8.2   Standards for Heterogeneous Networks

8.2.1   Bridging

One way to create heterogeneous networks is to use bridging. Bridging is standardized by the IEEE 802.1 working group (802.1D, 802.1Q, and 802.1AC). It enables several different links to create a single IP network, as shown in Figure 8.1. With bridging, it is possible to aggregate links of different technologies. This is what, for example, PLC/Wi-Fi or Ethernet/Wi-Fi extenders do. It is also possible to connect two logical networks (e.g., two logical PLC networks, as described in Section 7.3). Typically, most heterogeneous bridges are connected to only two technologies, and all frames of one technology are directly transmitted to the other technology, but it is also possible to have a bridge with more than two technologies.

A bridge maps the destination MAC address of a packet with an outgoing interface, based on a forwarding table (or forwarding information base) that creates such a mapping. This mapping is static, which means that the bridge has no flexibility to choose on which technology the packet will be sent.

8.2.2   Heterogeneous Networks at the IP Layer

Some devices (e.g., a hybrid router) have several interfaces and expose one IP address per interface. In this case, the decision of which technologies to use is thus made at the IP layer, and the source station has virtually no information on which technology to choose. In practice, the choice of the IP address, and thus of the destination interface, is often static (i.e., based on a predetermined order). Again, this solution typically does not provide much flexibility in heterogeneous networks. Each application can typically be assigned to a different static interface.

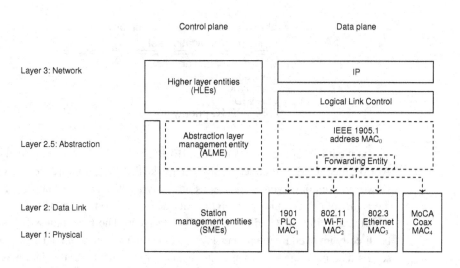

Figure 8.2 IEEE 1905.1 layers for the control plane and the data plane. The elements added by IEEE 1905.1 are in dashed lines. IEEE 1905.1-2013. Adapted and reprinted with permission from IEEE. © IEEE 2013. All rights reserved.

8.2.3   Overview of IEEE 1905.1

The IEEE 1905.1 standard introduces an abstraction layer (AL) between the MAC and IP layers, as illustrated by Figure 8.2. This means that the logical link control (LLC) layer, which provides the interface between the IP (network) and MAC (medium access) layers, does not need to know which technologies are present, or to decide which technology to use. Instead, the abstraction layer of IEEE 1905.1 exposes its own 48-bit MAC address, different from all other MAC addresses of the network, thus acting as a unique, common interface that hides the underlying technologies. Compared to the bridge that must use the outgoing interface corresponding to the specific MAC address of the packet, a 1905.1 device receiving a packet with the 1905.1 MAC address as destination can send it to any outgoing interface. The forwarding entity is responsible for choosing which technology to use, and can do so on a per-packet basis, which offers flexibility. The forwarding entity is not defined by the standard and is implementation dependent. However, it must be interoperable with the IEEE 802.1 bridging standards. We discuss the forwarding entity further in Section 8.2.4.

To manage the different interfaces, IEEE 1905.1 introduces an abstraction layer management entity (ALME). IEEE standards conceptually rely on so-called management entities; the physical layer management entity (PLME) and the MAC layer management entity (MLME) provide interfaces through layer-management functions that may be invoked (these functions either obtain or set the parameters of the layers, e.g., for Wi-Fi, such functions might be invoked to obtain or set the current channel of operation, the SSID, etc.). The station management entity (SME) is a layer-independent management entity that can invoke functions of different layers (PLME and MLME). There is one SME per technology. The higher-layer entities (HLEs) represent the man-

Table 8.1 Classification set for forwarding rules

Fields of classification set
MAC destination address
MAC source address
EtherType
VLAN ID
IEEE 802.1Q priority code point (PCP) field

agement entities of higher layers. With IEEE 1905.1, the HLEs can directly invoke the ALME; the ALME is then responsible to invoke the SMEs of the different technologies. However, the convergence layer does not fully replace each technology. In particular, the SMEs of each technology might continue to send control or management messages with the specific MAC addresses of the interfaces. The technology SMEs might also be directly invoked by the higher-layer entities; in particular, the HLEs might decide to send a data message on to a specific technology. In this case, the data message is sent with the MAC addresses of the technology.

8.2.4   Forwarding Entity

The forwarding entity can decide to send different frames from different flows simultaneously to different technologies, thus enabling high throughput gains due to multiplexing – to switch technology in the middle of a flow, for example, if a new station connects on one technology and causes higher interference, or in case of link failure. To this end, the forwarding entity uses forwarding rules: a forwarding rule associates a list of interfaces to a classification set (see Table 8.1), and all frames matching the classification set are forwarded to the corresponding interfaces. The classification set includes the source and destination MAC addresses, the Ethertype, but also the priority code point (PCP), a field of the 802.1Q header (VLAN tagging) that defines different classes of QoS (best effort, video, voice, etc.). Forwarding rules are set through the abstraction layer management entity (ALME).

The IEEE 1905.1 standard also defines a topology discovery protocol (see Section 8.4), which enables the IEEE 1905.1 devices to gain visibility on the network conditions, so that the forwarding entity can make accurate decisions based on the up-to-date topology.

8.3   Technologies

The IEEE 1905.1-13 standard supports the following technologies: PLC (IEEE 1901), Wi-Fi in the sub-GHz band (802.11af), in the 2.4 GHz and 5 GHz bands (IEEE 802.11a, 802.11b, 802.11g, 802.11n, 802.11ac), 60 GHz Wi-Fi (IEEE 802.11ad), Fast Ethernet and Gigabit Ethernet (IEEE 802.3u and 802.3ab), and coaxial cables (MoCAv1.1). Other technologies can be supported in the future by the IEEE 1905.1 standard, since the standard defines a two-octet MediaType field with specific values for each supported technology (see Table 8.2).

Table 8.2 IEEE 1905.1 media types

octet 1	octet 0	Medium
0 (Ethernet)	0	IEEE 802.3u Fast Ethernet
0	1	IEEE 802.3ab Gigabit Ethernet
0	2 to 255	reserved values
1 (Wi-Fi)	0	IEEE 802.11b (2.4 GHz)
1	1	IEEE 802.11g (2.4 GHz)
1	2	IEEE 802.11a (5 GHz)
1	3	IEEE 802.11n (2.4 GHz)
1	4	IEEE 802.11n (5 GHz)
1	5	IEEE 802.11ac (5 GHz)
1	6	IEEE 802.11ad (60 GHz)
1	7	IEEE 802.11af (< 1 GHz)
1	8 to 255	reserved values
2 (PLC)	0	IEEE 1901 wavelet
2	1	IEEE 1901 FFT
2	2 to 255	reserved
3 (MoCA)	0	MoCA v1.1
3	1 to 255	reserved values
4 to 254	0 to 255	reserved values

As mentioned earlier in this book, Wi-Fi is the most popular solution due to the explosion of mobile applications. It is usually used in the 2.4 GHz band (802.11b, 802.11g, 802.11n, and 802.11ax) and the 5 GHz band (IEEE 802.11a, 802.11n, 802.11ac, and 802.11ax). The other technologies usually act as an extender for Wi-Fi. The main reasons for the existence of these extenders are the connectivity demands and problems caused by Wi-Fi interference or bad channel conditions. In recent years, multichannel Wi-Fi has been widely employed, with multiple Wi-Fi devices simultaneously utilizing channels in the 2.4 and 5 GHz bands. Residential and enterprise APs typically support both 2.4 and 5 GHz bands and can load balance the number of clients between the two bands.

A Wi-Fi standard exists also in the sub-GHz band (between 54 and 790 MHz), 802.11af. Because the frequencies are lower, the range is larger, with up to 1 km, with a maximum PHY rate of 35.6 Mbps.

Another wireless technology compatible with the IEEE 1905.1 standard is 802.11ad, which works in the 60 GHz band. This technology offers throughput up to 7 Gbps, but it has a much smaller range of at most 10 m. Signal is also strongly attenuated by obstacles (walls, bodies, etc.) and often requires line-of-sight communication. This also means that spatial reuse is possible, which makes it a great candidate for certain applications.

MoCA is also a popular technology of IEEE 1905.1. Communication over coaxial cables achieves 1.4 Gbps data rates, using bands in the frequency range of 500–1650 MHz and with 100 MHz bandwidth. The main advantage of this technology is high reliability with packet error probability of about 10^{-6}. The interference is also limited within a local network, as shielded coaxial wires prevent interference from neighboring households or other technologies. MoCA applications mainly focus on multiroom digital video recorders, over-the-top streaming content, gaming, and ultra HD streaming.

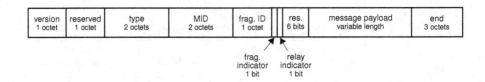

Figure 8.3 IEEE 1905.1 CMDU format.

Although it provides high reliability and rates, MoCA requires an already established coaxial system and possibly more than two coaxial plugs, which are limited in today's households.

Finally, Ethernet communication provides the highest reliability, but it can be the most costly. It is usually deployed in enterprise buildings with a sophisticated network structure. In residential environments, users often employ Ethernet for nonmobile applications, where the main disadvantage is the requirement for new wiring.

8.4   Topology Discovery with IEEE 1905.1

The IEEE 1905.1 standard describes a protocol that enables the devices to discover the network topology, that is, to discover neighboring 1905.1 devices and IEEE 802.1 bridges, and to exchange various link metrics. Discovering the network topology can then be used by the forwarding entity to choose to which technology to transmit a packet.

8.4.1   Control Message Data Unit (CMDU)

The IEEE 1905.1 standard describes the control message data units (CMDUs) that are used for the exchange of information between the abstraction layers of different 1905.1 devices. These messages are encapsulated in an Ethernet frame with a specific Ether-Type 0x893A. The MAC addresses of the CMDU Ethernet header might either be the 1905.1 MAC address or any of the underlying technologies' MAC addresses.

The format of a CMDU is shown in Figure 8.3. It contains a one-octet version (currently only 0x00); a reserved field of one octet currently unused; a message type, whose possible values are shown in Table 8.3 and described in more detail in Sections 8.4.2, 8.4.3, and 8.5; a message identifier (MID) of two octets that identifies each message; a fragment identifier (FID) of one octet that identifies each fragment of a fragmented message; one octet that serves to indicate if the message is the last fragment (fragment indicator, 0 or 1 on bit 7 of the octet) and if the message has been relayed (relay indicator, 0 or 1 for bit 6), the remaining 6 bits being currently unused; a variable-length payload, with different type-length-value sequences (TLVs); and a three-octet end-of-message TLV of value 0x000000.

In Table 8.3 which describes the CMDU types, transmission type "multicast" means that the relay indicator is set to 0, and a 1905.1 device that receives this CMDU does

Table 8.3 IEEE 1905.1 CMDU types

Field value	Message type	Transmission type
0x0000	Topology discovery	Multicast
0x0001	Topology notification	Relayed multicast
0x0002	Topology query	Unicast
0x0003	Topology response	Unicast
0x0004	Vendor specific message	Any
0x0005	Link metric query	Unicast
0x0006	Link metric response	Unicast
0x0007	AP autoconfig search	Relayed multicast
0x0008	AP autoconfig response	Unicast
0x0009	AP autoconfig Wi-Fi simple config	Unicast
0x000A	AP autoconfig renew	Relayed multicast
0x000B	Push button event notification	Relayed multicast
0x000C	Push button event join notification	Relayed multicast

Source. IEEE 1905.1-2013. Adapted and reprinted with permission from IEEE. © IEEE 2013. All rights reserved.

not retransmit it. Transmission type "relayed multicast" means that the relay indicator is set to 1, and a 1905.1 device that receives this CMDU retransmits it on all interfaces on which it has not yet received and retransmitted it. "Unicast" means that the message is addressed to a specific station (identified by its IEEE 1905.1 or technology-specific MAC address).

8.4.2 Neighbor Discovery

A 1905.1 device sends a "topology discovery" multicast CMDU on all its interfaces (i.e., on all technologies to which it is connected) once every minute. This means that each 1905.1 device knows the other neighboring 1905.1 devices. If a 1905.1 device detects a change in the topology, it must send a "topology notification" message that is relayed to all devices. This notification message does not contain any topology information; it only contains the 1905.1 MAC address of the device. To retrieve topology information, a 1905.1 device can send unicast "topology query" messages to another 1905.1 device. This 1905.1 device will respond with a unicast "topology response" that contains its capabilities (the technologies it is able to use) as well as the neighboring devices it has found (1905.1 devices or non-1905.1 devices, such as 802.1 bridges).

To discover 802.1 bridges present in the network, each 1905.1 must send a multicast IEEE 802.1 bridge discovery message along with each topology discovery message. The IEEE 802.1 bridge discovery message is a link layer discovery protocol (LLDP, as defined by IEEE 802.1AB) multicast message, sent with the 802.1 LLDP EtherType 0x88CC. An IEEE 802.1 LLDP-capable bridge will retransmit this message with its own MAC address; an IEEE 802.1 bridge that is not LLDP capable will not retransmit this message. This enables a 1905.1 device to detect the presence of 802.1 bridges on a link between two 1905.1 devices.

- If a 1905.1 device receives a 1905.1 topology discovery message and an LLDP message with the same source MAC address, it knows that there is no 802.1 bridge on this 1905.1 link.

- If it receives the two messages but with different MAC addresses, it knows that there is (at least) one LLDP-capable 802.1 bridge on this 1905.1 link.
- If it receives only a 1905.1 topology discovery message, it knows that there is (at least) one non-LLDP-capable 802.1 bridge on this 1905.1 link.

8.4.3   Link Metrics

The protocol described above enables each 1905.1 device to discover the topology of the network, that is, which links exist for the different technologies. In addition, the IEEE 1905.1 standard introduces link-metric query messages, with which each 1905.1 device can query another 1905.1 device to get link metrics for one or all neighbors. It can get the link metrics for either the transmitter or receiver side of the link. A link-metrics response for the transmitter side contains the following fields:

- estimated number of lost packets on the transmit side of the link during the measurement period (any integer, four octets);
- estimated number of packets transmitted by the transmitter of the link on the same measurement period used to estimate lost packets (any integer, four octets);
- the maximum MAC throughput of the link estimated at the transmitter and expressed in Mbps, a function of the PHY rate and of the MAC overhead (any integer, two octets);
- the estimated average percentage of time that the link is available for data transmissions (integer value between 0 and 100, two octets); and
- the PHY rate estimated at the transmitter of the link expressed in Mbps (any integer, two octets).

A link-metrics response for the receiver side contains the following fields:

- estimated number of lost packets on the transmit side of the link during the measurement period (any integer, four octets);
- estimated number of packets transmitted by the transmitter of the link on the same measurement period used to estimate lost packets (any integer, four octets); and
- the received signal strength indicator (RSSI) in dB, if the link is on Wi-Fi (any integer, one octet).

8.5   Security and Authentication in IEEE 1905.1

One key feature of the IEEE 1905.1 is that it describes unified security mechanisms that enable several 1905.1 devices (including the clients) to easily form a secure network. It provides three authentication modes: a mandatory push-button configuration (PBC) and two nonmandatory configurations: user-configured passphrase/key (UCPK) and a near-field communication network key (NFCNK). 1905.1 provides a unified framework and builds on the security mechanisms of each 1905.1-compliant technology.

8.5.1   User-Configured Passphrase/Key

In this mode, a global 1905.1 passphrase must be defined by the user for the 1905.1 network. Then, the IEEE 1905.1 standard defines how to use this passphrase to compute a key for each of the security mechanisms of the different technologies: WPA/WPA2 for Wi-Fi, NMK or PWK for PLC (see Chapter 7), MoCA 1.1 privacy password, and so on. Each key for each technology is computed through a hash of the 1905.1 passphrase and of a pre-defined message different for each technology. This means that if one key for one technology is compromised, the 1905.1 passphrase or the keys for the other technologies are not compromised.

8.5.2   Push Button Configuration

The push-button configuration enables a 1905.1 device to join a network by pressing a physical or logical button on the joining 1905.1 device, and on any other 1905.1 device already in the network. The two 1905.1 devices do not have to be connected to the same technology. The PBC mode is similar to the simple connect mode described in Chapter 7 and to the Wi-Fi protected setup (WPS) mode.

When a user pushes the physical or logical PBC button of a 1905.1 device, this device triggers the equivalent PBC procedure for each underlying 1905.1 technology to which it is connected (e.g., for PLC, it would trigger the procedure described in Section 7.3.1).

8.5.3   Near-Field Communication Network Key

In this mode, the 1905.1 key is copied between a key carrying device (KCD), for example, a near-field-communication-capable smartphone, and other near-field-communication-capable 1905.1 devices with different technologies. As described in Section 8.5.1, the 1905.1 key is then used to build the key for the different underlying technologies. Compromising the key of one technology does not compromise the keys of the other technologies.

8.5.4   802.11 AP-Autoconfiguration

The 1905.1 standard defines a protocol for automated configuration of a secondary 802.11 access point (Wi-Fi AP). The primary Wi-Fi AP is called the registrar; when a 1905.1-compatible Wi-Fi AP is successfully authenticated to the 1905.1 network (using either of the procedures described in Sections 8.5.1, 8.5.2, and 8.5.3) but has an unconfigured 802.11 interface, an AP autoconfiguration process is automatically triggered. This process consists of two phases, as follows.

Registrar discovery phase: The unconfigured AP sends an AP-autoconfiguration search CMDU (see Table 8.3) that contains the frequency band of the unconfigured AP requesting autoconfiguration (2.4 GHz, 5 GHz, or 60 GHz). If the APs support several bands, the whole process is repeated for each band. This search CMDU is a relayed multicast message, which means that it will eventually reach the registrar AP, potentially through multiple hops with non-Wi-Fi transmissions. Upon reception, the registrar sends a unicast AP autoconfiguration response CMDU, if it supports the frequency band contained in the search CMDU.

802.11 parameter configuration phase: Once the unconfigured AP has received the AP-autoconfiguration response CMDU, the two APs exchange AP-autoconfiguration Wi-Fi simple configuration (WSC) CMDUs. The WSC contains information like authentication type, encryption type, or BSSID. If the registrar AP's parameters change, the registrar sends a relayed multicast AP-autoconfiguration renew CMDU. Upon reception of this message, all AP enrollees start again the parameter-reconfiguration phase in order always to keep the up-to-date parameters.

8.6   Routing and Path Selection

The IEEE 1905.1 standard does not define any rule for the forwarding entity, which means that the routing protocols that are to be used in IEEE 1905.1-compliant networks are implementation dependent. In this section, we describe some routing protocols that have been proposed for heterogeneous networks. The best protocol to use naturally depends on the performance optimization goal (highest throughput, lowest latency, highest reliability, etc.). In Sections 8.6.1 and 8.6.2.1, we focus on throughput, which is typically the most challenging in heterogeneous networks and what users are typically the most interested in optimizing.

8.6.1   Heterogeneous Networks and Multipath Routing

In the last few years, multipath routing, which splits traffic between two or more paths, has gained popularity as a way to improve performance. The most popular solution is implemented at the transport layer: Multipath TCP (MPTCP) enables a TCP flow to be split between two paths [2, 3]. MPTCP has been widely studied in single-hop heterogeneous networks with Wi-Fi and cellular, and it has been shown to improve performance in general, and in particular reliability [4], throughput [5, 6], and latency [7]. MPTCP has been widely adopted, in particular, by several major vendors: all Apple operating systems (for smartphones and computers) now support MPTCP and an MPTCP implementation exists for all Linux-based devices [8]. MPTCP is not the only multipath solution recently developed: multipath solutions that support UDP traffic are also being considered (e.g., multipath QUIC [9][1]).

Multipath solutions have the potential to bring significant performance improvements in heterogeneous home networks, in particular in terms of throughput. Let us consider an example where a desktop computer is connected to a hybrid Wi-Fi/PLC access point, as depicted in Figure 8.4. If the desktop computer and the router use only Wi-Fi to

---

[1] QUIC (Quick UDP Internet connections) is an encrypted-by-default Internet transport protocol that provides a number of improvements to accelerate HTTP traffic and to make it more secure, with the intended goal of replacing TCP on the web.

communicate, they can accommodate up to 60 Mbps. However, if they use the two technologies simultaneously and because PLC and Wi-Fi do not interfere with each other, they can accommodate up to 110 Mbps, that is, an improvement of more than 80 percent.

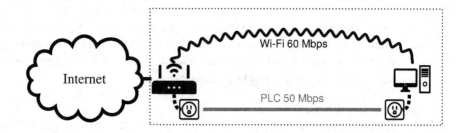

Figure 8.4 Scenario with a PLC/Wi-Fi router, and a PLC/Wi-Fi desktop computer.

These performance improvements are not limited to single-hop paths. Let us consider a second example, depicted in Figure 8.5, with a laptop connected to the hybrid PLC/Wi-Fi access point through a PLC/Wi-Fi extender. The connection between the laptop and the PLC/Wi-Fi extender is Wi-Fi only. Because PLC and Wi-Fi do not interfere with each other, 20 Mbps can be sent over the hybrid PLC-Wi-Fi Path 1. This is what happens with a typical PLC/Wi-Fi extender. But this consumes only one-third of Wi-Fi resources (the traffic is sent at a 20 Mbps rate on a 60 Mbps link). To improve throughput, some traffic can be sent over the two-hop Wi-Fi Path 2. The amount of traffic needs to be carefully calibrated and depends on the characteristics of the two Wi-Fi links A–B and B–C. These links cannot transmit simultaneously and need to share the capacity because Wi-Fi is a shared-medium technology. In this simple case, a computation on the remaining two-thirds of Wi-Fi resources can give us the rate x to send over Path 2 as the solution of $x/30 + x/60 = 2/3$, i.e., $x \simeq 13.2$ Mbps. Hence, by using both Path 1 and Path 2, it is possible to achieve a 66 percent improvement, compared to using Path 1 alone. However, choosing carefully the amount of traffic to send on each path is not easy, and we discuss this question in Section 8.6.4.

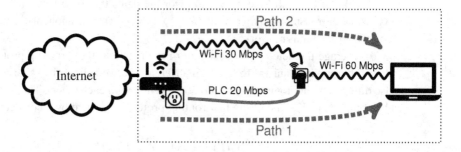

Figure 8.5 Scenario with a PLC/Wi-Fi router, a PLC/ Wi-Fi range extender, and a Wi-Fi user (laptop).

If we consider the example of Figure 8.4, it is easy to see that in terms of through-put, the optimal number of paths to use is equal to the number of technologies: the optimal throughput is reached when the two technologies are used. If the router and the desktop computer are connected, in addition to the Wi-Fi and PLC links, through a MoCA connection, then the optimal throughput is reached with three paths (one path per technology). This result can be extended to multihop networks [10]: in small net-works like home networks (in which all links of one technology typically interfere with each other), the optimal number of paths is equal to the number of technologies that are available between two devices. This shows that heterogeneous networks technologies and multipath routing are tightly linked together.

8.6.2   Single-Path and Multipath Routing Protocols for Heterogeneous Networks

In this section, we describe some routing protocols that have been described for hetero-geneous networks.

8.6.2.1   Single-Path Routing Protocols

There is a vast literature on single-path routing protocols for heterogeneous networks. They have been mostly designed for multichannel ad hoc wireless networks, but are applicable to be used with all heterogeneous networks. The general idea is the same for all these protocols: they define a weight function W on the paths, and use the path mini-mizing this function. This can be done with source routing (the path is fully determined at the source) or hop-by-hop routing (every intermediate station decides the next hop).

Most existing weight functions designed for routing protocols in heterogeneous net-works consist of two components: one component W_L that depends only on the links included in the path, and that aims at favoring paths that use good links (i.e., with high link capacity), and one component W_I that aims at favoring paths that yield low inter-ference – since interference reduces the end-to-end throughput. The total weight of one path P is computed by $W(P) = W_L(P) + W_I(P)$. These protocols start by defining a metric w on the links of each technology, typically a function of the link capacity c_l: for example, the expected transmission time (ETT), proportional to $1/c_l$. Most of the routing protocols use the link metrics summed over the links of the path, as the com-ponent W_L of the path function W, to favor the paths that have faster links (i.e., links with smaller transmission times): $W_L(P) = \sum_{l \in P} w(l)$. This component can be applied to a nonheterogeneous network.

To leverage the heterogeneous nature of the network, these routing protocols intro-duce different mechanisms to reduce interference between links, by favoring paths that use different technologies (as they do not interfere with each other). This is captured by the component W_I. One of the first routing protocols for heterogeneous networks [11] uses

$$W_I(P) = \max_{\text{tech. } t} \sum_{l \in P \text{ using } t} ETT(l), \tag{8.1}$$

which represents the bottleneck technology, that is, the technology that has the highest total ETT. However, this protocol is not isotonic. Isotonicity is defined as follows. A weight function W is isotonic if, for any path P_1, P_2, P_3, P_4, $W(P_1) \leq W(P_2)$ implies both $W(P_1 \oplus P_3) \leq W(P_1 \oplus P_3)$ and $W(P_4 \oplus P_1) \leq W(P_4 \oplus P_2)$; that is the concatenation (indicated by the symbol $\oplus$) of a same path P_3 or P_4 does not change the order in terms of W for P_1 and P_2. This condition is important because it has been proven [12] that isotonicity is a necessary and sufficient condition for Dijkstra's and Bellman–Ford shortest-path algorithms to converge to the path minimizing W or, in the case of hop-by-hop routing, for paths to be guaranteed loop-free, which means that finding the best path is more computationally intensive. Also, if the weight function is not isotonic, only source routing (and not per hop) can be used.

Another way to favor paths with different technologies is to use a channel-switching cost (CSC) as the component W_I. The CSC is an extra weight assigned to each node u of the network, equal to some value $w_s(u)$ when a path switches interfaces at node u (index s in w is for switching), and to $w_{ns}(u)$ when a path does not switch interfaces (index ns in w is for no switching). The weight of a path is the sum of the weights of its links (W_L) and of the CSCs of its intermediate nodes. Requiring $w_s(u) < w_{ns}(u)$ at each node u favors paths with alternating technologies. For example, one can use $w_s(u) = 0$ (switching interface does not cause any interference between links). If a path uses two links with the same technology, the end-to-end throughput is very roughly divided by 2 (this is the case if the two links have the same capacity; if one link has smaller capacity, the end-to-end throughput will be divided by a factor between 1 and 2). For this reason, it makes sense to use a value for $w_{ns}(u)$ of the same order of magnitude of $w(l)$; for example, one can use $w_{ns}(u) = \min_{l \text{ out of } u} w(l)$. It is possible to make the CSC isotonic by performing the shortest-path computations on the virtual graph of the network interfaces [13].

This accounts for intraflow interference, i.e., for the impact of interference between links of a same path. Other metrics consider in addition interflow interference, i.e., they try to favor links that are unlikely to be subject to interference due to other flows. This is typically taken into account in the W_L component. For example, IRU uses $w(l) = N(l) \cdot ETT(l)$ [13], where $N(l)$ is the number of neighboring nodes that would interfere with link l; CATT sums ETT over all interfering links [14]. iAware uses $w(l) = ETT(l)/IR(l)$ [15], where $IR(l)$ is the current interference ratio on link l. Note that to be practical and avoid path instability, the protocols that use interflow interference typically make the assumption that all neighboring links are used. When the number of flows is small, this can be an impractical assumption, and using a protocol that does not consider interflow interference yields better results [1].

8.6.2.2   Multipath Routing Protocols

Most of the literature studying multipath routing in multichannel wireless networks was working in the context of large ad hoc networks. In this context, the most efficient strategy is to avoid interference altogether, which is why these works try to build maximally disjoint paths ([e.g., 16–20]), that is, paths that impact each other as little as possible. Home networks, however, are usually small, and all links of a technology are very likely to interfere with each other. In this context, when trying to optimize throughput, maximally disjoint paths are necessarily paths that each use one and only one technology. But this choice is not optimal in terms of throughput; for example, in Figure 8.6, which represents the same network as Figure 8.5 with an additional PLC link between stations B and C, using the two Wi-Fi-only and PLC-only paths yields a total throughput of approximately 26.6 Mbps, whereas using Path 1 and Path 2 as in Figure 8.5 yields a throughput of about 33.2 Mbps.

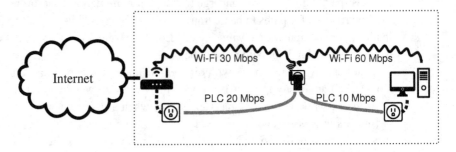

Figure 8.6 Scenario with a PLC/Wi-Fi router, a PLC/Wi-Fi range extender, and a Wi-Fi user (desktop computer).

One protocol that aims at finding the multipath that maximizes throughput has been described in [1]. It runs a single-path procedure to find the N best paths and then evaluates the achievable throughput when these paths are used simultaneously. It has been shown to be efficient in finding the best multipath in terms of throughput. However, it requires source routing, and its efficiency comes at the cost of an increased complexity, which makes it suitable for small networks like home networks, but less so for large enterprise networks.

8.6.3   Multipath Routing and IEEE 1905

In this section, we study how multipath routing can be used in IEEE 1905.1 networks. We are interested in using multiple paths within the local network. To use multiple paths outside the local network, multipath solutions at the transport layer, such as MPTCP, are required.

The IEEE 1905.1 standard describes the forwarding entity, responsible for choosing which technology to which to transmit the packet. As described in Section 8.2.4, the forwarding entity follows forwarding rules, that match the outgoing interfaces to a classification set. Several outgoing interfaces can be associated to a classification set. However, a frame that matches the classification set must be forwarded to all outgoing interfaces with which it is associated. This means that for a given flow (for which all frames share the same classification set), multipath can only be used by replicating the frames on the two paths.

8.6.3.1   **Reliability and Latency**

Multipath can be used for reliability and latency improvements: if the packets are repeated on two paths and the duplicates are removed at the receiver, then losses or delayed packets on only one path have no impact, as the packets will continue to arrive from the other path. To improve reliability and latency, maximally disjoint paths [e.g., 16–20] are preferred (with the example of Figure 8.6, it would be better to use two paths that use different links, such as the Wi-Fi-only path and the PLC-only path; if an additional hybrid extender is present in the network, it would be better to use it for one of the two paths).

8.6.3.2   Throughput

Using multipath to improve throughput is possible in IEEE 1905.1 when there are several flows. For example, in Figure 8.5, if the laptop is simultaneously streaming video and downloading a file or using VoIP, or if it is watching multiple streams, the different flows can be marked with different PCP fields. In this case, the forwarding entity is able to use two different paths for the different flows, thus enabling the multiplexing gains that heterogeneous networks can offer. The IEEE 1905.1 standard does not specify any routing protocol; routing protocols are implementation dependent and would depend on the characteristics of the network and of the flow (how many nodes, how many flows, what performance metric to optimize, etc.). IEEE 1905.1 leaves space for vendor-specific 1905.1 CMDUs, which would enable 1905.1 devices to exchange 1905.1 messages in order to agree on a routing protocol, such as one of those described in Section 8.6.2.1. In addition, the standard enables the nodes to have knowledge of the network topology, including of the link metrics (see Section 8.4), which makes it possible to easily use complex routing protocols that are not only based on hop count. Most existing routing protocols use link metrics that are specified by IEEE 1905.1 (e.g., link capacity, airtime, packet errors).

8.6.4   Congestion Control

It has been shown that saturating multihop paths is inefficient and can lead to congestion collapse with packet losses and instabilities [21, 22]. This is also true with multipath routing [1], as links from different paths might interfere with each other. Many congestion control algorithms have been designed for heterogeneous networks [e.g., 1, 23–30]. These algorithms are powerful but would usually be complex to use in practice: Some of them are not compatible with current protocols, some of them take a long time to converge, and some of them require information that is not easy to access (e.g., queue lengths).

Because of this complexity, the most heavily used algorithm for congestion control in today's networks remains TCP. TCP does not cohabit smoothly with multipath used on a single flow, as it reacts to packet drops: packet drops on one path will impact the whole flow, even though the other path might be underloaded. This was the reason for the design of multipath TCP [2, 3]. Multipath TCP is, however, not well suited for multipath limited to local networks, especially with IEEE 1905.1, since the primary goal of IEEE 1905.1 is to hide the underlying technologies and to abstract the different technologies as one unique link. Thus using multipath at the transport layer is not adapted to IEEE 1905.1.

As we have seen, IEEE 1905.1 does not enable a device to use multipath with a single flow; multiplexing the traffic on multiple paths is possible only for multiple flows. This is a limitation, as this does enable a user to enjoy the benefits of multipath when it uses a single flow. But it also makes the design of multiplexing different flows with multiple paths easier, because employing TCP on each of these flows is sufficient, if congestion control is required.

8.7   Summary

In this chapter, we have presented the IEEE 1905.1 standard, designed for the interoperation of different technologies in heterogeneous networks. We have discussed the different mechanisms (topology discovery, security, etc.) defined by this standard. This standard mostly provides unifying frameworks. The specific features and functions of the underlying technologies are preserved. We have also discussed the question of routing and congestion control in heterogeneous networks, in particular in compliance with the IEEE 1905.1 standard. Since IEEE 1905.1 is not compatible with single-flow multipath, complex multipath congestion-control algorithms are not required in IEEE 1905.1 networks. However, in heterogeneous networks, multipath routing offers high performance gains (in terms of throughput, latency, reliability) when multiple flows are present.

References

[1] S. Henri, C. Vlachou, J. Herzen, and P. Thiran, "EMPoWER hybrid networks: Exploiting multiple paths over wireless and electrical mediums," in *ACM International Conference on Emerging Networking EXperiments and Technologies (CoNEXT)*, 2016, pp. 51–65.

[2] A. Ford, C. Raiciu, M. Handley, S. Barré, and J. Iyengar, "Architectural guidelines for multipath TCP development," *RFC 6182*, 2011.

[3] A. Ford, C. Raiciu, M. Handley, and O. Bonaventure, "TCP extensions for multipath operation with multiple addresses," *RFC 6824*, 2013.

[4] I. Lopez, M. Aguado, C. Pinedo, and E. Jacob, "SCADA systems in the railway domain: Enhancing reliability through redundant multipath TCP," in *2015 IEEE 18th International Conference on Intelligent Transportation Systems (ITSC)*, pp. 2305–2310.

[5] C. Raiciu, C. Paasch, S. Barré, A. Ford, M. Honda, F. Duchêne, O. Bonaventure, and M. Handley, "How hard can it be? Designing and implementing a deployable multipath TCP," in *USENIX Conference on Networked System Design and Implementation (NSDI)*, 2012, pp. 399–412.

[6] Y.-C. Chen, Y.-S. Lim, R. J. Gibbens, E. M. Nahum, R. Khalili, and D. Towsley, "A measurement-based study of multipath TCP performance over wireless networks," in *ACM Internet Measurement Conference (IMC) 2013*, pp. 455–468.

[7] A. Frömmgen, T. Erbshäußer, A. Buchmann, T. Zimmermann, and K. Wehrle, "ReMP TCP: Low latency multipath TCP," in *2016 IEEE International Conference on Communications (ICC)*, pp. 1–7.

[8] C. Paasch and S. Barré, "Multipath TCP in the linux kernel," [Online]. Available: www.multipath-tcp.org. [Accessed: November 6, 2019].

[9] Q. De Coninck and O. Bonaventure, "Multipath QUIC: Design and evaluation," in *ACM International Conference on Emerging Networking EXperiments and Technologies (CoNEXT)*, 2017, pp. 160–166.

[10] S. Henri and P. Thiran, "Optimal number of paths with multipath routing in hybrid networks," in *IEEE International Symposium on a World of Wireless, Mobile and Multimedia Networks (WoWMoM)*, 2018, pp. 1–7.

[11] R. Draves, J. Padhye, and B. Zill, "Routing in multi-radio, multi-hop wireless mesh networks," in *ACM 10th Annual International Conference on Mobile Computing and Networking (MobiCom)*, 2004, pp. 114–128.

[12] V. Borges, D. Pereira, M. Curado, and E. Monteiro, "Routing metric for interference and channel diversity in multi-radio wireless mesh networks," in *AdHocNow*, 2009, pp. 55–68.

[13] Y. Yang, J. Wang, and R. Kravets, "Interference-aware load balancing for multihop wireless networks," Department of Computer Science, University of Illinois at Urbana-Champaign, Tech. Rep., 2005.

[14] M. Genetzakis and V. Siris, "A contention-aware routing metric for multi-rate multi-radio mesh networks," in *IEEE Conference on Sensor and Ad Hoc Communications and Networks (SECON)*, 2008, pp. 242–250.

[15] A. P. Subramanian and M. Buddhikot, "Interference aware routing in multi-radio wireless mesh networks," in *IEEE Workshop on Wireless Mesh Networks*, 2006, pp. 55–63.

[16] P. Dong, H. Qian, K. Zhou, W. Lu, and S. Lan, "A maximally radio-disjoint geographic multipath routing protocol for MANET," *Annals of Telecommunications*, vol. 70, no. 5–6, pp. 207–220, 2015.

[17] J.-Y. Teo, Y. Ha, and C.-K. Tham, "Interference-minimized multipath routing with congestion control in wireless sensor network for high-rate streaming," *IEEE Transaction on Mobile Computing*, vol. 7, no. 9, pp. 1124–1137, 2008.

[18] J. Gálvez, P. Ruiz, and A. Skarmeta, "Multipath routing with spatial separation in wireless multi-hop networks without location information," *ACM Computer Networks*, vol. 55, no. 3, pp. 583–599, 2011.

[19] S.-J. Lee and M. Gerla, "Split multipath routing with maximally disjoint paths in ad-hoc networks," in *2001 IEEE International Conference on Communications (ICC)*, vol. 10, pp. 3201–3205.

[20] B. Yan and H. Gharavi, "Multi-path multi-channel routing protocols," in *IEEE International Symposium on Network Computing and Applications (NCA)*, 2006, pp. 27–31.

[21] V. Gambiroza, B. Sadeghi, and E. Knightly, "End-to-end performance and fairness in multihop wireless backhaul networks," in *ACM 10th Annual International Conference on Mobile Computing and Networking (MobiCom)*, 2004, pp. 287–301.

238   **Heterogeneous Networks and IEEE 1905**

[22] R. Srikant, *The Mathematics of Internet Congestion Control*. Springer Science & Business Media, 2004.

[23] S. Henri, C. Vlachou, and P. Thiran, "Multi-armed bandit in action: Optimizing performance in dynamic hybrid networks," *IEEE/ACM Transactions on Networking*, vol. 26, no. 4, pp. 1879–1892, 2018.

[24] H. Mohsenian-Rad and V. Wong, "Joint optimal channel assignment and congestion control for multi-channel wireless mesh networks," in *2006 IEEE International Conference on Communications (ICC)*, pp. 1984–1989.

[25] A. Giannoulis, T. Salonidis, and E. Knightly, "Congestion control and channel assignment in multi-radio wireless mesh networks," in *IEEE Conference on Sensor, Mesh and Ad Hoc Communications and Networks (SECON)*, 2008, pp. 350–358.

[26] B. Radunović, C. Gkantsidis, D. Gunawardena, and P. Key, "Horizon: Balancing TCP over multiple paths in wireless mesh network," in *ACM 14th Annual International Conference on Mobile Computing and Networking (MobiCom)*, 2008, pp. 247–258.

[27] X. Lin and S. Rasool, "A distributed joint channel-assignment, scheduling and routing algorithm for multi-channel ad-hoc wireless networks," in *IEEE INFOCOM International Conference on Computer Communications*, 2007, pp. 1118–1126.

[28] W.-H. Tam and Y.-C. Tseng, "Joint multi-channel link layer and multipath routing design for wireless mesh networks," in *IEEE INFOCOM International Conference on Computer Communications*, 2007, pp. 2081–2089.

[29] L. Zhou, X. Wang, W. Tu, G.-M. Muntean, and B. Geller, "Distributed scheduling scheme for video streaming over multi-channel multi-radio multi-hop wireless networks," *IEEE Journal on Selected Areas in Communications (JSAC)*, vol. 28, no. 3, pp. 409–419, 2010.

[30] M. Alicherry, R. Bhatia, and L. E. Li, "Joint channel assignment and routing for throughput optimization in multi-radio wireless mesh networks," in *ACM 11th Annual International Conference on Mobile Computing and Networking (MobiCom)*, 2005, pp. 58–72.

9 Conclusion

9.1 Summary of Book Content

In this book, we have explored residential and enterprise power line networking, mainly focusing on the IEEE 1901 standard and its practical implementations. PLC can provide high capacity and offers easy installation. For these reasons, PLC is often combined with Wi-Fi into hybrid networks that extend coverage while offering mobility. Hybrid networks, which can employ additional technologies to PLC and Wi-Fi, are specified by the IEEE 1905.1 standard, which is discussed in Chapter 8.

We have presented several PHY/MAC-layer features of PLC, which strongly affect the performance. PLC channel modeling has unique features due to the noise on the power line and the electrical appliances attached. Owing to these unique features, it has different spatial and temporal channel-quality variability compared to Wi-Fi. Thus PLC and Wi-Fi can complement each other, yielding performance benefits for hybrid networks. The PHY layer design of PLC addresses different types of noise, in particular the one periodic to the mains. PLC has a more complex PHY design than Wi-Fi, enabling adaptive modulation per OFDM carrier and addressing a higher degree of frequency selectivity compared to the wireless medium.

We have studied the MAC layer of PLC, which is similar but more complex than that of Wi-Fi. PLC can be more efficient but less fair than Wi-Fi, depending on its CSMA/CA protocol configuration. We have provided a performance evaluation and models of multiple configurations that can be used to optimize the performance for different quality of service requirements, either minimizing latency or maximizing throughput.

We have also presented the security features and the network management of PLC. The central coordinator is similar to a Wi-Fi AP and manages the PLC network. It ensures security of data, as well as efficient performance. Contrary to common Wi-Fi infrastructures that use a star topology with all stations communicating to AP only, PLC enables peer-to-peer communication, that is, a mesh topology. The PHY layer of PLC and the MAC-level security features create different vulnerability levels compared to Wi-Fi, as discussed in Chapter 7. If tone map adaptation is used, PLC frames are more secure than Wi-Fi on the PHY layer, as the intruder has to be able to decode the packet with the specific tone map. It is recommended that all critical unencrypted information be transmitted via per-link adapted tone maps and not ROBO modes with PLC.

Our last chapter focuses on hybrid networks and their standardization. We have discussed the technologies involved, and how to route traffic through a hybrid network.

Chapter 8 describes the standardized link metrics, while Chapter 5 provides guidelines and methods for their estimation in PLC. Hybrid devices are very popular in residential networks where security and data privacy are important. Chapter 8 discusses the new security functions introduced in IEEE 1905.1.

9.2   Technology Limitations

We have identified several key performance benefits of PLC. Nevertheless, there are still open problems that challenge large-scale deployments of PLC. First, as we discuss in Chapter 5, PLC spatial variability is challenging to predict. Link performance highly depends on the infrastructure wiring, electrical appliances attached, and electrical network structure. Second, PLC links can be asymmetric, hence expected performance can vary significantly even between the two directions of a single link. Contrary to Wi-Fi, PLC path loss models are often very complex and need to take into account the characteristics of electrical appliances attached to the grid. There is not much research or standardization on path loss models for different PLC environments and wiring, which would be beneficial for designing enterprise PLC in large scale.

Another limitation of PLC research is on the network topology, which is currently a mesh. The advantage of this topology is that every station can communicate with each other, without the need for message passing through the CCo. This yields high efficiency in residential environments, where, typically devices need to communicate with each other. However, it can lead to disadvantages in enterprise environments or access networks where communication is flowing from a single source to multiple destinations into a star topology. The mesh versus star topologies define different applications in PLC networking, and so far, PLC standardizations have mostly considered the mesh mode for indoor networks. Star topology in PLC is challenging, as temporal capacity variation might worsen link quality and exclude stations from the topology, if the CCo is static. This is an effect that happens with Wi-Fi too under mobility: in such a case, the user roams to a different AP, which equivalently could be a different CCo in PLC. Roaming between CCos and logical networks has not been discussed in the standard and would be necessary for large-scale star deployments. To advance the technology further and have more optimal scheduling in star topology, the CCo could also implement centralized scheduling, similarly to IEEE 802.11ax Wi-Fi APs.

Finally, compared to Wi-Fi, PLC does not use channelization, which helps scale the network in large enterprise and residential buildings. To avoid hidden terminals and minimize interference, Wi-Fi uses different channels by dividing the available unlicensed spectrum into 20/40/80/160-MHz channels. So far, no PLC technology has implemented channels, and all of them use the whole available bandwidth. Therefore any logical networks on the same or adjacent electrical grids can interfere with each other.

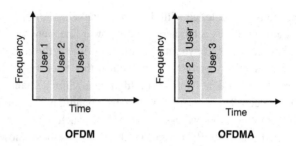

Figure 9.1 OFDM versus OFDMA protocols for scheduling users.

9.3   Looking Ahead

Based on the limitations discussed above, there are many interesting research areas for next-generation PLC. The generalization of attenuation and noise models in different environments is crucial for the future evolution and deployment of PLC, especially in enterprise or industrial environments. The investigation of different PLC topologies, such as mesh versus star, and as a result, CCo roaming, is also an interesting topic for future networks.

A star topology could enable new MAC-layer scheduling techniques with PLC. Following Wi-Fi IEEE 802.11ax paradigms, PLC networks could deploy orthogonal frequency-division multiple access (OFDMA). OFDMA divides the bandwidth between users in smaller "resource units" that can be an arbitrary or fixed number of subcarriers, enabling simultaneous downlink or uplink communication to multiple users, as we can see in Figure 9.1. It is used by LTE networks and recent IEEE 802.11ax Wi-Fi networks. OFDMA resource allocation has also been proposed for indoor HomePlug networks and access networks with relays [1, 2]. Given the frequency selectivity of PLC channels, the CCo can use the tone map and channel estimation information to optimally assign subcarriers to users based on their channel, in order to maximize throughput. This can be implemented in a mesh topology by preallocating subcarriers in beacons or in a star topology that enables dynamic low-overhead allocation, where the CCo has the most up-to-date buffer information per client and can allocate resources given channel and buffer status reports. This solution has not been implemented or standardized yet, potentially due to the complexity of the software and hardware required. This can be addressed by new simulations, standardization, and research engineering efforts.

Alternatively to a complex OFDMA design, PLC can deploy channelization in next-generation standards, in order to scale deployments. Similarly to Wi-Fi, PLC logical networks could divide the bandwidth into a few channels, and the network administrator or a channel allocation algorithm at the CCo could determine the channel per network. The advantages would be a collision-free environment and elimination of hidden terminals. The disadvantages of channelization would be a drop in PHY rate proportional to the bandwidth, and bandwidth wastage in the case of unused channels. This can be potentially tackled with a dynamic channel allocation instead of static network configurations. A challenge with dynamic channelization is that it would require the CCos to be able to sense each other over the power line, hence a thorough topology planning that takes into account dynamic channel conditions throughout the day would be needed.

The spatial variability discussed in this book might be a challenge for large-scale broadband PLC networks with high bandwidth demands. However, it would challenge less GreenPHY networks with narrowband devices, such as IoT and smart metering. Ensuring coverage is less challenging in such networks, as they use specific low rates (e.g., the three ROBO ones of IEEE 1901), and these known tone maps make the channel modeling and capacity planning more tractable. The challenge with such networks then comes from the MAC layer and the scalability issues with a large number of smart meters or sensors. Another challenge would be security and privacy, as using a known ROBO modulation scheme enables an intruder to decode frames on the PHY without needing any tone map information. It is interesting to model the traffic requirements and the CSMA/CA performance when the GreenPHY network scales.

Hybrid networks also have open research areas. In addition to IEEE 1905.1 technologies, more candidates can be combined with PLC, such as different wireless standards that offer mobility. In particular, PLC combined with visible light communications (VLC) has been investigated by Song et al. [3]. The authors present different system models with the goal of reducing the network complexity, and they build a prototype testbed showing practical performance results. New hybrid-network technologies and system models that guarantee wide coverage, reliability, and/or high rates are key factors in next-generation PLC deployments and standardizations.

References

[1] S.-G. Yoon, D. Kang, and S. Bahk, "OFDMA CSMA/CA protocol for power line communication," in *2010 IEEE International Symposium on Power Line Communications and Its Applications (ISPLC)*, pp. 297–302.

[2] Q. Zhu, Z. Chen, and X. He, "Resource allocation for relay-based OFDMA power line communication system," *MDPI Electronics*, vol. 8, no. 2, p. 125, 2019.

[3] J. Song, W. Ding, F. Yang, H. Yang, B. Yu, and H. Zhang, "An indoor broadband broadcasting system based on PLC and VLC," *IEEE Transactions on Broadcasting*, vol. 61, no. 2, pp. 299–308, 2015.

Index

Printed in the United States
by Baker & Taylor Publisher Services